Protein Machines, Technology, and the Nature of the Future

Wyatt Galusky

Protein Machines, Technology, and the Nature of the Future

palgrave
macmillan

Wyatt Galusky
SUNY Morrisville
Morrisville, NY, USA

ISBN 978-3-031-08716-5   ISBN 978-3-031-08717-2   (eBook)
https://doi.org/10.1007/978-3-031-08717-2

This Palgrave Macmillan imprint is published by the registered company Springer Nature Switzerland AG.
The registered company address is: Gewerbestrasse 11, 6330 Cham, Switzerland

To Ramona and her sisters

Acknowledgements

My work on this book results from innumerable conversations, interactions, and opportunities afforded by others. I owe debts of gratitude to so many that it would be a fool's errand to try and settle them all here. I do want to thank the anonymous reviewers who offered meaningful comments on how to improve the book. I would also like to offer a special thanks to the landlords in VA, who were generous enough to allow us to raise chickens, even when we had no idea what we were doing. I would also like to acknowledge my STS colleagues, who indulged many a talk I delivered about chickens. I'm especially indebted to Benjamin R. Cohen, who has always been a sounding board for me, and a willing and meaningful collaborator, as well as Chris Henke, who kept asking me to speak with his Food class, to my benefit. In the same spirit, I would like to thank my students in New York, especially those at Hamilton College and SUNY Morrisville, who helped be work out these ideas across multiple semesters. I of course need to thank all of my chicken friends, to whom I did my best, though invariably never good enough. And my family, who ended up on their own chicken journeys. Finally, I can't really capture the fullness of the gratitude I have for my spouse, Kelly Ann Nugent, who encouraged my attempts, helped to mourn my losses, and never failed to support my newest efforts to learn and to do.

CONTENTS

CHAPTER 1

An Introduction: Protein Machines with Flaws

Abstract The purpose of the book is to use human efforts to control chickens and ourselves in order to reimagine human relationships to nature (non-human animals, humans, environments, and agricultural practice). I identify key historical elements of our technological engagements with chickens to confront the stakes involved in configuring these "protein machines" now and in the future, in order to deepen our understanding of just what the stakes are when we make chickens part of our technological world, where those stakes come from, and what kinds of future we might pursue if we want to mitigate the ecologically unsustainable ways bound together with the modern technological chicken.

Keywords Protein machine • Technology • Chickens • Sustainability

This is a book about protein machines. We should probably explore this a bit. A protein machine is a concept that refers to many things, depending on the point of emphasis. On the one hand, it can mean machines *made out* of protein. This version of the term relates to bodies—in the context of agriculture and food production, we might think of non-human animal bodies like cows, pigs, chickens, or sheep. This configuration can also be applied to humans, though we'd rarely do so in the same context. Humans are of the same stuff as non-human animals, after all, but not normally considered food.[1] On the other hand, we might instead think of machines

W. Galusky, *Protein Machines, Technology, and the Nature of the Future*, https://doi.org/10.1007/978-3-031-08717-2_1

designed to *produce* protein. This second version connects more readily to systems—technological and agricultural systems designed to produce protein for consumption. The fact that most animal protein in our contemporary world is produced in industrialized contexts reinforces the notion that we get our protein from integrated systems linking animals, feed, ecosystems, economies, and patterns of work—it is, in fact, a machine that produces *a lot* of protein.

These two versions of protein machine, of course, are not mutually exclusive, especially to the extent that bodies (animal and human) are necessary components of systems (networks of producers, distributors, and consumers) that generate the desired output, and that these animal bodies can be engineered to produce this protein more efficiently. That is part of the story of this book—about how animal bodies (in the first sense) and food systems (in the second) become intertwined and shape each other. How, historically, those interconnections have become more numerous, more prosperous, and more troubling. So, to say that this is a book about protein machines is to say that we will be analyzing both types of those machines and how they are interrelated.

The troubling aspect is where the "with flaws" part of the title for this introduction comes in, especially given where these relationships between bodies and systems break down. The full concept, protein machines with flaws, comes from a quote in Michael Pollan's 2006 book, *The Omnivore's Dilemma*. In a section discussing Joel Salatin's pig-raising techniques and philosophy, Pollan notes that Salatin treats a pig as a pig—a whole animal that has desires and needs not simply restricted to the production of protein—rather than "a protein machine with flaws."[2] The presumption here is that, for conventional producers, animals are viewed primarily as entities that produce protein. And they could do their "jobs" better if they did not get sick or bored and, as a result, expend energy on activities other than adding to their stores of muscle and the protein that we eat. These so-called flaws at the level of the individual are magnified within the contemporary food system in the aggregate, in terms of inefficiencies (feed conversion, life span, illness), suffering (pain, maladaptive behaviors, disease), and outputs (waste streams, human disease, political action). So, it's not just the individual animal that can been seen as flawed, but rather the entire system and its configuration. After all, it is not the waste stream of a single pig, but the cumulative output of a pig population that might rival a small city, for example, that generates hazards for local communities. Importantly for our analysis here, both places where we might locate these flaws, in bodies or in the systems that contain those bodies, we have also sought to *fix* those flaws through a variety of means: to make the body more efficient, to make the system more efficient, to produce more protein more quickly at a lower economic or environmental cost. As a result, in addition to understanding just what our protein machines are, this book also asks, what methods have been developed or promoted to make them better? In particular, what do these purported solutions tell us about our relationships to nature, our visions of the future, and the sustainability of that future?

The protein that will form the center of our inquiry into these various machines is chicken. My focus on chicken is not just because the word neatly if confusingly designates both the animal and the meat many people now consume. It is also because of chicken's meteoric rise in consumption as a part of the American diet since the middle part of the twentieth century. This consumption gained traction after World War II, as chicken transformed into "meat," rather than a less desirable alternative to it. This transformation was profound, as chicken became something more people in the United States chose to consume, and even became a preference for an everyday meal, overtaking beef in popularity by the early part of the twenty-first century.

It is also because chicken, in a very tangible sense, has shaped a large part of my adult life, despite the fact that I do not come from a family of poultry producers. Consider the following historical moments that were not part of a plan, but helped to form the contours of my chicken-mediated existence:

- A teenager gets his first job, at the age of sixteen, at a brand-new branch of Kentucky Fried Chicken, which opens in a growing suburb of Dallas, Texas. The job is desirable as it pays US$0.35 over the minimum wage at the time.[3] At orientation, he receives some of the perks of the job, including a navy-blue ascot cap, a folder bearing the declaration, in bold red letters, "You Are KFC" (which the adult version of the teen still possesses), and the knowledge that he would eat for free while on the clock. He is not warned about how many pairs of shoes he will go through, due to the corrosive work conditions like an eternally greasy floor. Leftover chicken at the end of the night can be taken home, but his interest quickly wanes—nothing like being surrounded by chicken for 6 hour shifts to lose one's appetite for it. Instead, because such disaffection is rampant in the entire staff, leftovers are often just thrown away. He stays at that job for a calendar year, before turning in the uniform (but keeping the folder).

- After finishing a degree in Environmental Ethics and having spent those years talking about duties to animals in the abstract (with no material effect), a young adult takes a summer job teaching conversational English in South Korea, and decides to ride a bus to visit a market that opens at midnight in Seoul. For much of the 90-minute journey, the bus shares the four-lane highway with a truck delivering live chickens to the city. The bed of the truck is stacked five high with cages, which are in turn stacked with chickens. The cages are so full that a few of their inhabitants meet their fate earlier than intended, being crushed to death or suffocated by the density. His window is next to the truck the entire journey. This experience, more visceral than the innumerable discussions he's had about the fate of farm animals,[4] prompts him to start eating a mostly vegetarian diet.

- A not so young adult rents a farmhouse in Virginia, which has an old poultry house on the property. He decides to try his hand raising eight Red Star Sex-Linked hens in those existing facilities. The chickens spot the flaws in the structure (the gaps in the boards, the holes in the floor) before he does. Predators do, too. The chickens take to roosting in the trees rather than be too easily picked off in a building that keeps them in, but not other animals out. After losing most of the original flock, he builds a smaller, more secure structure to better accommodate the sole remaining bird. Now flockless, she improvises and spends her days with two male goats as companions, before succumbing to disease. All told, the flock is gone within 18 months, the first seven from predation, the last from illness. The next time he wants eggs, he goes to the supermarket.

Each of these moments, taken as anecdotal experiences of an individual (me), do not reflect anything of particular significance. They do reflect a certain privilege (the capacity to choose one job that pays slightly better than others, to work overseas, to raise animals as a hobby rather than a necessity), but are not extraordinary. To call them somewhat ordinary, in fact, would be the point, as these experiences can also be viewed as partially representative of shifts in the place of chicken within American culture over the course of the later stages of the twentieth century.

As historians such as Horowitz and Boyd[5] have noted, the chicken became a much more prominent part of the American diet after World

War II, as poultry producers and distributors worked to expand both the body of, and the market for, the chicken as an everyday food item. Chicken moved from a meal used to mark a special occasion, or something eaten only after little else was available, to a staple of the American diet. In the same historical frame of reference that witnessed chicken's rise in "meat" status, alterations in agricultural production and processing, spurred in part by subsidies and other government incentives to produce agricultural commodities at ever accelerating abundance, helped to supply an ever-expanding array of fast food franchises. Fast-food purveyors like Kentucky Fried Chicken expanded to occupy a stable place within the casual dining firmament, such that the chain could open a new restaurant in the sub-urbs—like the one that employed me—and offer free meals to its employ-ees. Chicken became cheap, plentiful, and desirable. No longer a chicken in every pot, but a KFC in every town.[6] These franchises helped, in turn, to transform food culture in the United States, while providing low-skilled, close to minimum wage employment opportunities to a growing American workforce.

Greater industrialization of agriculture more generally, coupled with a more global food supply, freed certain people in industrialized countries from privation because of a lack of food. Focusing on quantity over qual-ity, U.S. agricultural policy promoted a more is better mentality that spilled over into consumption. In addition, the "fast fooding" of food ways brought with it a slew of human and environmental health problems, and precipitated heightened awareness of the human, animal, and envi-ronmental costs associated with cheap, processed foods for those who could afford to have a choice. The possibility to choose to eat as an expres-sion of values and of identity not moored to culture came to exist, as well. Couple this level of freedom with a growing visibility of agricultural prac-tice, and people begin to express themselves even more particularly in part through their diets, in a context of choice and abundance. Individual diets can be a means of expressing one's politics, ethics, or social affiliations.

Responses to agricultural conditions also led to smaller scale attention to growing food, including the growth of a kind of hobbyist food culture, where people grew their own food in the context of abundance. Importantly, these gardens and hobby farms are pursued within the con-text of the industrial system—there to buffer any failure that occurs in backyard husbandry.[7] All of those elements—the growing presence of meat as a cheap and readily available foodstuff, the globalization of food production facilitated by increasing concentrations of animals, the growing awareness of and reactions to agricultural production facilitated by increasing concentrations of animals, the growing awareness of and reactions to agricultural productions by consumers—have led to complicated contemporary relationships to meat. It is a relationship torqued by equally fraught relationships to technology—reflected in increasing capacities to modify the world to suit specific human ends, awareness of the unintended consequences stemming from those transformations, and a general reliance on better technology to fix those problems. And I participated in many of those trends, from the age of fast food, to its inevitable backlash, as well as a chicken-mediated version of an insulated, small scale form of food production.

These experiences help shape the current analysis for the book. They do not, however, mark me out as an expert in poultry production per se, nor do they encompass anywhere close to the possible chicken-related perspectives one might have. I would invite the reader to help make a list of other possibilities that may reflect their own experiences. I would offer this non-exhaustive start:

- The person whose economic well-being is directly tied to the production of meat protein (such as the grower, the processor, or the corporate employee at big chicken producers).
- The person who, because of persistent economic and social inequities, confront questions of whether to eat, rather than what.
- The person who engages with food as nothing more than a means of achieving other goals, rather than a specific focal point of political or ethical reflection.[8]
- The person who eats no animal products at all, for reasons of health, politics, or ethics.

But while my own experiences are not representative of the totality of interactions with chickens, I do not want to deemphasize them. They matter, at least to me, in that they have focused my attention and they have enabled a kind of life. That is, my interactions with chickens have shaped my interests and have been shaped by my material circumstances. Each reader will have their own experiences and their own stories, which may amount to a single sentence (I have eaten chicken) or might fill an entire book.[9] What is important is that these stories matter, because to think about chicken, about chickens, about the food systems that help to produce them, about protein machines of these types, is to think about

ourselves. We become bound to, and implicated by, the creation and functioning of these protein machines in ways that are easy to overlook, yet
central to our own existence. And as we struggle to understand that inter-
dependence, and the kinds of responsibilities that it creates, we also have
to reckon with our own identities as protein machines with flaws.

Two Contests/Two Chickens

The first two chapters of the book are largely organized around two stories
about chicken, related to the staging of two contests, separated by almost
60 years, each of which aimed at trying to improve the chicken—1948s
Chicken of Tomorrow Contest & a 2008 *in vitro* chicken meat contest.
The idea of what counts as improvement for each event, given their disparate historical and cultural contexts, varies quite dramatically—whereas
one contest sought to expand the market demand for chicken meat, the
other looked to fill an established demand more humanely. But both rely
on technological innovations in producing animal protein to help accom-
plish their respective tasks. I encountered these contests because of the
experiences I had raising my own small flock of eight birds that I men-
tioned above. I was in the mind of improvement, because my efforts had
mainly been failures—chickens roosting in trees, laying eggs in mysterious
places, getting eaten, getting sick. So many chickens behaved unexpectedly and ultimately died before their time (or at least related to my very
specific expectations) that I was forced to confront my own inadequacies
in this area. But, maybe, per these contests, the faults were not mine as
much as the chickens', and what I needed were better ones.

So, how do we make a better chicken? The 1948 Chicken of Tomorrow
Contest, sponsored by the A&P Foodstores, posed just such a question,
and offered US$5000[10] in prize money to the entrant who could best
answer it. The problem at that time, for contest organizers, was that not
enough people were eating chicken regularly, as it tended to be pricey,
undesirable, or not part of a wider food culture. As such, for the contest
the concept of "better" was already carefully prescribed—submissions
were judged on a set of criteria reflecting a variety of traits, such as lower
production costs, quicker growing times, more uniform bodies, and fuller
breast meat, which might lead to a more desirable food. The contest's aim
was to find just such an appealing breed of meat-type[11] chicken which,
after processing, would look the like a wax model of an ideal, platonic
chicken much like one would now find in a local supermarket.[12]

It was a beauty pageant of sorts where "dressed" takes on a very different connotation. While we're at it, so does "winner," which in this case was an entry by the Vantress Hatchery and Poultry Farm of Marysville, California—a cross between the Cornish and New Hampshire breeds of chicken. The goal of the contest, then, was to make chicken protein better (more economical, more desirable) by making a better chicken. This overall process engaged both parts of the contest's name, by shaping the physiology and ecology of the bird, but also by shaping "tomorrow." A better chicken led, it was hoped, to a bigger market, and would help transform the United States into a nation of chicken-eaters. Transform it, it did—leading to increased consumption of chicken, up to the point where chicken overtook beef as the most consumed protein in the US in 2011.

The chicken (one form of protein machine) embedded within food systems (another form) became so efficient at producing protein that market-determined prices dropped and chicken became ubiquitous in supermarkets and in homes. As I will argue in Chap. 1 (A Chicken, Part I), the Chicken of Tomorrow Contest represents a kind of improvement through control—not just on the body of the chicken, but on the environment that conditions the animal's existence. The contest is a point on a trajectory of increased control as a means of generating increased output—chickens live in more conditioned environments, eat more engineered food, have more directed biologies. The contest, then, is a way of seeing how animal and ecology are part of the machine. In fact, the ecology is replaced by the human designed systems that seek to exert control. This story of a chicken, though, is also one of failure—of places where systems of control are not total, break down, and reveal areas where the responsibilities we have taken on become burdens too large to carry. These failures, in fact, lead to our second contest.

About 60 years later, another kind of contest was held that also sought to make better chicken.[13] This 2008 challenge, lacking the snappy name of the earlier version (which, frankly, was a real missed opportunity, potential trademark issues aside), was sponsored by the People for the Ethical Treatment of Animals (PETA). For PETA, the fundamental problem was not too little consumption of chicken or too small a market, but rather too many chickens—how do we deal with people's appetites for chicken, while lessening the suffering of the animals needed to meet them? The charge, with the winning entry offered a US$1 million prize, invited people to produce chicken protein through *in vitro* means. This method relies on laboratory-based techniques for generating meat through the use of muscle cells alone—chicken without the chicken. Like the contest from the middle of the last century, this one sought an alternative to the more commonplace bird of the era in order to make meat better (more precise, more controlled). Unlike the Chicken of Tomorrow contest, however, this campaign sought to make the protein differently, in order to make it appear the same—to cater to a pre-existing market that had a demonstrated desire to eat chicken. That is, undertaken in a context of an existing nation of chicken eaters, the PETA-backed contest focused on using the *in vitro* production methods in order to establish an equivalent output compared with current chicken production techniques, in both taste and market viability. But while the due date for a viable entry to the contest has come and gone with no winning entry, the *in vitro* technology still remains a vital part of a new imagined tomorrow for the contemporary world.

These two contests, separated by over half a century, appear to have little in common, except the protein. Both contests did inherit a specific cultural and technological world, and hoped to use the production of chicken protein as one vehicle to change it. Using chicken, to borrow a phrase, to invent the future.[14] It appears, however, that this is where any substantive similarities cease. One might find it easier to draw out contrasts:

- Promoting chicken consumption vs. tolerating it;
- Cultivating a chicken body to be a better protein machine vs. bypassing that body altogether;
- Staging a contest with multiple suitors and a clear path to success vs. simply announcing a prize employing a technology applied at an unproven scale.

These contrasts might be easy to identify but might also obscure a more fundamental connection. What underlies these disparate protein machines and why does it matter? As these contests move us from yesterday to today to tomorrow, from the field to the factory to the laboratory, are there commonalities that allow us to compare the solutions they offer and the forms of responsibility they demand? Thus, in Chap. 2 (A Chicken, Part II), I argue that this new chicken of tomorrow contest and the *in vitro* methods it promotes inherit more from what came before than might appear at first blush. The move from the factory to the laboratory packs up and brings along an intensified desire for control. The body of the chicken is not simply removed from the equation in favor of only producing muscle, but replaced by human-designed systems. And the burden of responsibility increases—in this sense, *in vitro* technologies do not substi-tute or replace the underlying orientation and motivations of the existing system, but rather magnify them. The struggle, then, will be for us to imagine a future that might contain this new kind of protein machine and one that might not.

THE SCIENCE OF SPECULATION: TWO FUTURES

Efforts to imagine a future that contains *in vitro* protein machines are important, though they are speculative—this technology contains "mul-tiple promissory narratives."[15] About what virtues it brings and what futures it offers. What might it mean to explore this type of responsibility and these kinds of relationships, or anticipate the kind of future that would contain them? Luckily for us, we have writers and other visionaries who have explored these possibilities since the early part of the last century. Not precisely in terms of the contemporary approaches to *in vitro* tissue generation, but certainly in terms of the idea of animal-less meat and chicken-less chicken. Going back into time for glimpses of an imagined future may allow us to get a better handle on the stakes and the possibili-ties associated with our contemporary choices.[16] We can trace curiosity into the possibility of lab-based chicken as far back as the spectacular (and dubious) work of Alexis Carrel, who claimed to have been able to keep alive, and grow in perpetuity, a piece of embryonic chicken heart in a suit-able medium as early as 1912. This work inaugurated what I have dubbed the age of expansion in fiction and other forms of speculation. Promoters such as Winston Churchill (who expressed his dreams for the future in a 1931 essay, "50 Years Hence"), as well as tongue-in-cheek skeptics such as the radio dramatist Arch Oboler (in 1937s radio drama, "Chicken Heart") and the authors Frederik Pohl and C.M. Kornbluth (featuring an organ-ism called Chicken Little, in the novel, *The Space Merchants*, published in 1951), reflected on a future where flesh can grow without limit[17]—be that limit skeletal or, at least in one case, geographic. When chicken flesh runs riot, however, the story tends not to end well. Later narratives shifted from dealing with nature's productivity unleashed to an engagement with biological plasticity harnessed and directed by human intention. Exemplified by Margaret Atwood's *MaddAddam* trilogy, what I designate the age of design tackles a world crafted by humans to contain what we want and omit what we don't—meat without animal suffering, for exam-ple. Atwood explores, in particular, what happens when this type of human-directed control appears possible—how far we might take it (into humans ourselves) and how illusory that sense of control really is (very).

Thus, in Chap. 3 (A Future, Part I), I argue that these fictional accounts of the future help us to map out the stakes in our current choices and technological explorations. Importantly, and collectively, these visions of the future correlate our creations with ourselves. A mass of flesh engineered to grow without limits reflects a society bent on consumption without end. A world fully designed by humans come to reflect humanity more fully, especially our flaws (short-sightedness, avarice, a certain lack of empathy). More than anything, these narratives allow us to pose the question—what kind of society/what version of humanity is reflected in this technology-mediated relationship to the world? By seeking to solve the problems of chicken through *in vitro* means, what and who are we seeking to sustain? These science fictional accounts each explore different ways in which the problems of meat have been "solved," and inform a society that would support such solutions.

We must then turn to try and answer those questions for ourselves, to stake a claim on the kind of sustainable future we might seek, situated in the present and projecting forward into an unknown. That is, we must go beyond this historical reconstruction and attempt to map out possible futures that will not simply unspool in front of us, but can be shaped by our own actions and intentions. One might speculate and generate counter narratives to the existing slate of more apocalyptic or celebratory fiction that takes a designed world as a given—these counter-narratives might either map out alternative visions of human/animal relationships (similar to what *Ecotopia* did with human-nature ones[18]) or at least insist, like characters in *Station Eleven* do regarding life in a world of collapse, that "survival is insufficient."[19] Instead, in Chap. 4 (A Future, Part II), I offer a method for exploring the states of these decisions and possible futures, by taking the idea of a protein machine, a technology that humans play a part in constructing, seriously. This is, if we embrace the idea of a protein machine, filtered through a more nuanced understanding of technology and the responsibilities that technology requires, we might at least be more prepared to confront a future more intentionally. It answers the question, what does it mean to think though chicken technologically? What, that is, might it mean to think about technology as a set of relationships and forms of responsibility? Borrowing from, among others, Martin Heidegger, Bruno Latour, and Peter-Paul Verbeek, the narrative looks at the specific ways in which humans and the natural world (specifically non-human animals and ecosystems) are co-constituted within networks of associations, and how such networks configure those relationships. What emerges is a renewed and sustained focus on *responsibility*, understood in terms of both creating and maintaining. These forms of being responsible have requirements that include knowledge systems and work, imposed on a humanity that would itself impose direction on the natural world. Responsibility is directly proportional to intention—the more humans attempt to express their intention on the natural world, the more they become responsible for creating and maintaining that world. Importantly, this relationship has multiple iterations, including backward into necessity (premised on systems that need to work as designed to enable the technology to function) and forward into possibility (mediating new ways of being based upon a functional technological system that can be taken for granted). In addition, responsibility is imprecise—human models of how the world works are restricted by limitations in the models themselves. Thus, I make the theoretical assumptions and approaches that have informed the earlier work more explicit and offer some possibilities regarding a new chicken of the future.

From a contemporary standpoint, meat does generate a number of problems, and those problems are exacerbated by specific human eating patterns and tastes. *In vitro* meat technology operates as a solution, to the extent that we take as given human dietary habits and the essential malleability of animal bodies and natural systems. In the context of human relationships to the world, then, we seek to maintain an orientation to the natural world, wherein animals become substances, and humans become fully responsible (in creation and in maintenance) for the nature that sustains them. This orientation, however, requires relationships and responsibilities which are founded on very tenuous assumptions, including the infallibility of human knowledge and the predictability of human energy systems. A more resilient approach to sustainability would include more, not less, proximity to animals and ecosystems (not substances), sharing the burdens of responsibility by establishing relationships that need not be fully controlled to be functional. The analysis illustrates elements of the human-animal-nature dynamic in terms of changing aspects of responsibility. These aspects allow us to get a fuller grasp of the ethical contours of the issues, not just in terms of how to solve problems, but also in terms of how we seek to relate to the world around us. Understanding the development of human relationships to these animals and ecologies, in the configuration of various kinds of protein machines, are key to confronting the kinds of future we wish to promote, and the kind of present we wish to sustain.

NOTES

1. Science fiction contains a lot of references to humans as machines made of protein, especially in terms of the human body that might be controlled by implants or other cyber-actuated means—William Gibson used the term "meat puppet," in *Neuromancer*. See Gibson, William. *Neuromancer*. New York: Ace Books, 1994. More to the point, Terry Bisson emphasized the protein-based human reality in his short story, "They're Made Out of Meat," in which alien visitors express their shock and dismay at the composition of sentient life forms on Earth. See Bisson, Terry. "They're Made Out of Meat," 1990. http://www.terrybisson.com/page6/page6.html. Accessed 19 March 2016. We will visit science fiction more formally later in the book.
2. Pollan, Michael. *The Omnivore's Dilemma: A Natural History of Four Meals*. New York: Penguin Press, 2006: 219.
3. Though I cannot be absolutely certain, my guess is that the pay represented the need to possess certain skills—namely, the ability to differentiate parts of the chicken (e.g., leg, wing, thigh, breast, and in KFC's case, a keel or center breast). These five different pieces were to be aggregated in various ways, depending on the size of the bucket or type of order. All white meat was extra.
4. "We know the facts, but we don't always realize them with that imaginative, emotional engagement that makes them vivid forces and deciding factors." Solnit, Rebecca. *The Faraway Nearby*. Reprint edition. New York, New York: Penguin Books, 2014: 151.
5. See Horowitz, Roger. "Making the Chicken of Tomorrow: Reworking Poultry as Commodities and as Creatures, 1945-1990." In *Industrializing Organisms: Introducing Evolutionary History*, edited by Susan R. Schrepfer and Philip Scranton, 215–35. New York: Routledge, 2004, and Boyd, William. "Making Meat: Science, Technology, and American Poultry Production." *Technology and Culture* 42, no. 4 (2001): 631–64.
6. Or, as Striffler puts it, "a chicken nugget in every fryer." See Striffler, Steve. *Chicken: the Dangerous Transformation of America's Favorite Food*. New Haven: Yale University Press, 2005.
7. Importantly, this context enabled people to make choices not based primarily on material or economic subsistence, but instead as an expression of identity. See, for example, Alexander, William. *The $64 Tomato: How One*

Man Nearly Lost His Sanity, Spent a Fortune, and Endured an Existential Crisis in the Quest for the Perfect Garden. Chapel Hill, NC: Algonquin Books of Chapel Hill, 2007.

8. As Gene Kahn, the founder of Cascadian Farm, told Michael Pollan, "This is just lunch for most people. Just lunch. We can call it sacred, we can talk about communion, but it's just lunch," *Omnivore's Dilemma*, 153.

9. Barringer, David. *Johnny Red: An Amatory Allegory*. Middletown, NJ: Word Riot Press, 2005, for example.

10. Adjusted for inflation, this equals roughly US$50,000 in 2015.

11. "Broilers referred not to a type of chicken, but to a stage of development that suited the animal to a certain type of cooking. Aggressive cross breeding of chickens following the war largely eliminated breeds outside of poultry fanciers, in favor of distinctions by form in which the animal would be used: layers, broilers, roasters, and so forth." Horowitz, Roger. *Putting Meat on the American Table: Taste, Technology, Transformation*. Baltimore: Johns Hopkins University Press, 2006: 104.

12. See Audio Productions. *Chicken of Tomorrow, The*, 1948. http://archive.org/details/Chickeno1948, for an example of this wax model, around the 1:48 mark.

13. The lack of article is intentional, as a means of signifying a change in focus away from the animal and toward the meat in isolation.

14. Virginia Tech, my alma mater, had a tagline "Invent the Future," from 2006 to 2017.

15. O'Riordan, Kate, Aristea Fotopoulou, and Neil Stephens. "The First Bite: Imaginaries, Promotional Public and the Laboratory Grown Burger." *Public Understanding of Science* 26, no. 2 (2017): 148–63. See also Stephens, Neil. "In Vitro Meat: Zombies on the Menu?" *SCRIPTed* 7, no. 2 (2010): 394–401.

16. See Belasco, Warren J. *Meals to Come: A History of the Future of Food*. Berkeley: University of California Press, 2006.

17. Carrel references what he calls "preventable occurrences" with regard to what puts a check on the growth of cells within a body. See Carrel, Alexis. "ON THE PERMANENT LIFE OF TISSUES OUTSIDE OF THE ORGANISM." *The Journal of Experimental Medicine* 15, no. 5 (May 1, 1912): 516–28.

18. See Callenbach, Ernest. *Ecotopia: The Notebooks and Reports of William Weston*. New York: Bantam Books, 1990.

19. See Mandel, Emily St. John. *Station Eleven*. New York: Alfred A. Knopf, 2016.

Bibliography

Alexander, William. *The $64 Tomato: How One Man Nearly Lost His Sanity, Spent a Fortune, and Endured an Existential Crisis in the Quest for the Perfect Garden.* Chapel Hill, NC: Algonquin Books of Chapel Hill, 2007.

Audio Productions. *Chicken of Tomorrow, The*, 1948. http://archive.org/details/Chickeno1948.

Barringer, David. *Johnny Red: An Amatory Allegory.* Middletown, NJ: Word Riot Press, 2005.

Belasco, Warren J. *Meals to Come: A History of the Future of Food.* Berkeley: University of California Press, 2006.

Bisson, Terry. "They're Made Out of Meat," 1990. http://www.terrybisson.com/page6/page6.html.

Boyd, William. "Making Meat: Science, Technology, and American Poultry Production." *Technology and Culture* 42, no. 4 (2001): 631–64.

Callenbach, Ernest. *Ecotopia: The Notebooks and Reports of William Weston.* New York: Bantam Books, 1990.

Carrel, Alexis. "ON THE PERMANENT LIFE OF TISSUES OUTSIDE OF THE ORGANISM." *The Journal of Experimental Medicine* 15, no. 5 (May 1, 1912): 516–28.

Gibson, William. *Neuromancer.* New York: Ace Books, 1994.

Horowitz, Roger. "Making the Chicken of Tomorrow: Reworking Poultry as Commodities and as Creatures, 1945–1990." In *Industrializing Organisms: Introducing Evolutionary History*, edited by Susan R. Schrepfer and Philip Scranton, 215–35. New York: Routledge, 2004.

———. *Putting Meat on the American Table: Taste, Technology, Transformation.* Baltimore: Johns Hopkins University Press, 2006.

Mandel, Emily St. John. *Station Eleven.* New York: Alfred A. Knopf, 2016.

O'Riordan, Kate, Aristea Fotopoulou, and Neil Stephens. "The First Bite: Imaginaries, Promotional Public and the Laboratory Grown Burger." *Public Understanding of Science* 26, no. 2 (2017): 148–63.

Pollan, Michael. *The Omnivore's Dilemma: A Natural History of Four Meals.* New York: Penguin Press, 2006.

Solnit, Rebecca. *The Faraway Nearby.* Reprint edition. New York, New York: Penguin Books, 2014.

Stephens, Neil. "In Vitro Meat: Zombies on the Menu?" *SCRIPTed* 7, no. 2 (2010): 394–401.

Striffler, Steve. *Chicken: The Dangerous Transformation of America's Favorite Food.* New Haven: Yale University Press, 2005.

CHAPTER 2

A Chicken, Part I

Abstract This chapter focuses on a type of protein machine, the chicken, linked to the Chicken of Tomorrow Contest from the 1940s. This contest sought to solve problems associated with producing chicken through technological interventions into the body of the chicken itself, and less obviously in the surrounding ecology. The chapter examines the historical trajectory of making chicken, as the chicken of tomorrow became the chicken of today and led to an array of problems: physiological, environmental, economic, epidemiological, ethical, and political. The chapter catalogues the problems that emerge with these historical developments. At the same time, I show how they are linked to a specific technological approach that makes assumptions about what animals are, who humans are, and how food systems should work.

Keywords Chicken of tomorrow • Factory farms • Industrialization • Food • Problems • Control

> *Follow the chicken and find the whole world.*
> —Donna Haraway, *The Companion Species Manifesto: Dogs, People, and Significant Otherness*, Chicago: Prickly Paradigm Press, 2003

W. Galusky, *Protein Machines, Technology, and the Nature of the Future*, https://doi.org/10.1007/978-3-031-08717-2_2

This chapter explores the process of making and remaking chicken, a process that involves the configuration of a protein machine that will serve those interests. In order to make a chicken of tomorrow in the middle part of the twentieth century, people had to construct two specific kinds of protein machines—a body and a system that were more efficient and more consistent than any that had come before. The components of those machines included a right animal body (a breed of chicken that, under the right set of circumstances, would exhibit the desired characteristics related to size, uniformity, and growth rate) and the right set of circumstances (a technological system designed to contain and promote that animal body, and including an environment, an economy, and a set of human relationships related to production and consumption). The purpose of the indefinite articles here and in the title of the chapter are meant to reinforce the fact that this machine and its machinery are historically contingent—not the result of an inexorable march toward a better future, but rather a specific vision of a kind of "future" and an idea of "better" which result in specific configurations of our food system that could have been otherwise. This vision included a transformed chicken (cheap and ubiquitous) and a transformed food culture—one that embraced chicken protein as a desired and regularly consumed food, rather than an unhappy compromise. And turning this vision into a reality came with certain requirements for environments and for people, and led to unplanned consequences for animals and for communities. That is, our protein machine has flaws that intersect a variety of components, including physiological, environmental, economic, epidemiological, ethical, and political aspects of our food system (here organized under the animal specific acronym of PEEEEP). Our task here will be to explore those elements that comprise this emergent protein machine—one that, by the later stages of the twentieth century, has completely transformed how people in the United States come to view and consume chicken. Disaffection with the chicken of their "today" led to the desire in the post-war U.S. to create a chicken of tomorrow, which requires a vision for the chicken and a vision for the future.

Configuring "Tomorrow" (The Factory)

About the time our story begins (the Chicken of Tomorrow contest first took place in 1948), my mother was growing up on a farm in rural Michigan. Her family kept chickens for eggs, and while they did consume the birds, such consumption was largely considered, as Roger Horowitz describes it, as a "substitute to meat"[1]—something to eat on a particular occasion, or as a canned winter protein when beef or pork wasn't to be had, but not as an everyday meal. The chickens that were consumed on my mother's family farm were either older laying hens no longer producing eggs in abundance or the occasional unruly rooster—basically, those chickens no longer adequately performing their prescribed tasks (either laying eggs or protecting the flock without being too much of a nuisance) were subject to becoming a meal. As a result, eating chicken was mostly the result of an unfortunate circumstance, though admittedly more so for the chicken. It certainly was not a meal that my mother's family looked forward to, though getting rid of an ornery, spur-clad rooster had its perks.

Of course, while more people lived on a farm in the middle of the last century than do now, not everyone did, nor did they kill their own animals for protein. Generally, outside of certain specialized markets, chicken existed on the periphery of meat consumption, a substitute when "real" meat was unavailable or needed to be given up. Instead, in the early twentieth century, chickens were often raised for eggs, often by women, for what was labeled by men in a derogatory fashion as "pin money."[2] Chicken raising did not remain on the periphery of the rural and agricultural economy, however, as people began to notice the economic possibilities of this more marginal element of farming—as Lawler found, quoting the 1933 *Progressive Farmer*, chicken raising has gone "from 'an unimportant farm chore' to 'a scientific business and major source of farm income.'"[3] This form of protein had economic potential, then, but still had to challenge the primacy of beef and pork in what we might term full-fledged "meat" status. It was not a given that people in general would come to perceive chicken as a desirable, everyday food, instead of an more exotic protein like rabbit or rattlesnake.

The period of World War II represented a potential turning point in the widespread consumption of chicken. During the war years, many people turned to it as a source of protein, as "real meat" was rationed because of the conflict in Europe. In fact, a civilian who raised and ate chickens was seen to be doing his or her part for the war effort.[4] As a result, the mid-1940s saw a spike in the consumption of chicken in the United States in response to the specific historical context. However, as Lawler notes, the looming return of these unrationed proteins posed a threat to the long-term maintenance of the market gains. What would happen when people could eat beef and pork again? Records show that there was a dip in the consumption of chicken after the end of the war.[5] A very real historical possibility existed that chicken would be perpetually relegated to a consolation protein in mainstream United States food ways.

For chicken producers, however, the wartime rise in consumption levels pointed to a more desired future—one where "chicken" became "meat" at least on par with beef and pork. But for that to happen, a series of transformations had to occur, related to the chicken and to the nation. Those transformations included bringing the price of chicken down, making the output more consistent, and making the eating of chicken more desirable. Or, more succinctly, changing the chicken, changing the process of chicken production, and changing the chicken consumer. The Chicken of Tomorrow contest played its part, helping to reimagine that potential future, by focusing on chicken breeds and processes more conducive to these long-range goals of producing a uniform, consistent, and cheap bird.

CREATING A PROTEIN MACHINE...

As we saw in the introduction, the Chicken of Tomorrow contest was designed to help stabilize demand for chicken, and then subsequently grow it, especially after the gains in the war years. The contest had an optimistic, promissory spirit, suggesting that if the right chicken breed could be found, then the future for poultry producers would be bright. And the obvious focus is on the bird—seeking to identify the breed that could literally embody all of the traits that the organizers prized. These were traits that connected to economic efficiency in one form or another. The chicken industry sought birds that converted feed to weight as efficiently as possible, had a high viability rate from birth to slaughter, and put on weight in all of the right places (prioritizing breast meat over others). The goal was to make chicken—as a protein—readily available, relatively inexpensive, and easily recognizable at the grocers, and in so doing entice people to include it as a regular part of their diets.

While it makes sense to focus on the chicken as the key element of this process and as a protein machine in its own right, we should also recognize that the more expansive concept of the machine needs to be considered. Without the larger system that helps to structure this new chicken, and the innovations that occurred there, as well, the kind of future that poultry producers envisioned would not come to pass. The components of this more expansive machine include innovations in various types of controls—engineered food, confinement structures, efficiencies in processing. These aspects condition the possibility of this chicken of tomorrow, and are important parts of the protein machine we are exploring. The chicken changes, but so does what it eats, where it lives and for how long, how that body moves through space (both while alive and as it goes through the various steps of production and processing), and how humans come into contact with that animal. This is how the machine is built to take us into our tomorrow.

(Component: An Animal Body)

To find the breed of chicken that would herald a brighter tomorrow for the industry, entries were solicited from breeders across the country. The Chicken of Tomorrow was not to be an individual bird, but rather an archetypal breed—one that was consistent enough to be created by, and processed in, the systems being designed around it. In an effort to facilitate the desired outcomes of this contest, the organizers devised experimental controls to ensure that as many extraneous variables as possible were accounted for. Producers shipped fertilized eggs to the contest site, where all entrants were collectively hatched (one metric being the percentage of viable eggs) and the chicks were raised under the same conditions—a specific engineered food, a relatively controlled environment—and then slaughtered after 12 weeks and 2 days.[6]

The chickens are evaluated throughout their three-month life, as well as after being dressed.[7] The contest used a series of checklists identifying qualities such as "factors effecting edible meat yield"[8] and the mortality rate at 12 weeks. The qualities promoted (proportionality, consistency, viability, size) were meant to produce the best chickens for growing new markets and increasing demand:

- Proportion—birds were sought that produced larger breasts and more white meat overall.
- Consistency—a breed was sought that had little variation from bird to bird, allowing for uniformity across dressed carcasses.
- Viability—more eggs hatched, and more chicks survived to slaughter.
- Size—the larger the bird at slaughter (based on feed rates and age), the cheaper the cost.

The ideal bird grew larger in its 114 days than its rivals, on the same amount of food, producing relatively uniform carcasses with more desirable meat, thus bringing down the food-to-weight ratio and, more importantly, the costs of production, all things being equal. And it was through this gauntlet that the Vantress Hatchery's Cornish-New Hampshire Cross passed to become the winner of the contest in 1948, with the Vantress chicken subsequently becoming a mainstay of the broiler industry.

The 1948 Chicken of Tomorrow Contest marked the beginning of a slow and steady rise in chicken consumption in the United States at the end of that decade into the twenty-first century.[9] The Contest is not the simple or direct cause of this rise, but it does reflect larger changes in the overall process of producing chicken,[10] including a desire to make uniform what "chicken" is for the consumer, by focusing on a bird that best embodies that ideal, achieved in part through selective breeding and, later, through a growing understanding of genetics. Because of their comparatively shorter growing times and relative cheapness, chickens were at the forefront of such genetic tinkering.[11] These efforts continue through the latter part of the century, leading to ever greater changes to chicken physiology[12] and to branding and intellectual property efforts by chicken producers.[13] As Schatzker notes, the Chicken of Tomorrow Contest never really ended.[14]

An important element of this process involves the restrictions placed on the chicken in terms of precisely where it fits within the surrounding context. That fit is conditioned on the chicken's capacity to produce protein, at the expense of other traits. That is, the chicken essentially becomes a smaller protein machine within the larger machinery. This ability to transform the chicken into a protein machine is achieved in part through a much more acute and detailed knowledge system related to protein production in that body. How to manipulate bodies into producing protein more quickly and efficiently requires new and precise knowledge of how the body works and how muscle production can be prioritized. Take, for example, the midcentury scientific discovery that antibiotics added to the feed of animals, even when they were not sick, promoted more body growth more quickly than those not fed the drugs.[15] No longer needing to devote energy to preventing illness, the bodies could devote that energy to producing muscle and fat. This example highlights how the protein machine relies on people who have learned how to tinker, who have become more aware of the chicken in its specialized role, and more able to match breeds as a means of maximizing yields. Ruth Harrison, in her book *Animal Machines* from 1964, puts it this way:

The factory farmer cannot rely, as did his forebears, on generations of experience gained from the animals themselves and handed down from father to son; he relies on a vast array of back-room boys with computing machines working to discover the breeds, feed, and environment most suited to convert food into flesh at the greatest possible speed, and every batch of animals reaching market is a sequel to another experiment or part of an experiment.[16]

Harrison captures here that kind of transition in the factory setting, related to how knowledge is acquired and by whom. Expertise won through experience has lost ground to those "back-room boys" who examine inputs and outputs—minimizing the former and maximizing the latter. There have been gains in efficiency, in the creation of a very specific kind of chicken, in part by creating new forms of scientific expertise that helps to enable that creation. That place in the machine for chickens, however, now also relies more heavily on how accurate and complete that knowledge is, so that the protein machine system (and the smaller protein machines within) continues to hum along. Harrison offers an important question by way of warning: "Does the agricultural student [...] know enough about the physiology of the animal and is his training wide enough?"[17] Now that the onus of defining chickens and producing chicken fall on the maintainers of the protein machine in this context, that question has more relevance than ever.

On the other hand, there is a kind of truncation of other forms of knowledge, where non-productive behaviors or traits are either ignored or sought to be corrected. These birds are made to fit within the machine, which requires that those doing the fitting have a very good sense of exactly how that can happen. This sense of fit is achieved not just by maximizing protein, but also through rooting out unproductive or inessential behaviors that are deemed worthless or detrimental to the function of this newly arranged machine. There is a specific way to be a chicken in this structure, and most other aspects do not count—behaviors like brooding[18] or establishing pecking orders or dust bathing. Within the logic of the system, we get what we want to call chicken, and everything else is designated as not-chicken. We take pains to enable this form of chicken, newly defined, to function in this attenuated sense. "Chicken" becomes a much more restricted concept.

The individual body of a chicken is not the only thing that gets simplified in this process—so, too, does the breed. The concept of breeds within the industry gradually fade—as Horowitz notes, by the 1950s and 60s, "[t]he bird had, in essence, been successfully standardized; it was a broiler, not a Plymouth Rock or Rhode Island Red."[19] Broiler becomes the specific manner in which chicken can be expressed within the system. Again, with the goal of achieving maximal protein yield from the individual chicken put through the process, broilers signify enough. Breed characteristics are basically irrelevant to the fundamental goal of producing protein. In order to understand more fully how that place for broilers was established, we should turn to trace the innovations and techniques in environment, feed, and processing that enable this particular configuration of the emerging protein machine.

(Component: An Environment)

Transformations of the environment that surrounding the chicken in order to turn it into a broiler emerged as a part of this newly configured protein machine. Enclosed structures allowed for the regulation of space, feed, water, and waste removal, erecting what begins to look more and more like a factory—chicks, water, food, and supplements operate as inputs, while waste and, eventually, protein emerge as outputs. And these structures would get bigger and bigger. For example, a mid-century leaflet entitled "Arkansas Broilers: A Forty Million Dollar Industry"[20] touts changes to the size of the houses and the related style of living for the birds. Brooder houses grew from a relatively modest 12 × 14 foot area to 40 × 400 (16,000 square feet). And, whereas with the smaller houses, "[i]t was common practice to turn the chicks out[side] as soon as weather permitted," the larger houses promoted chicks "kept in complete confinement." Current broiler houses in the US can be as large as 40,000 square feet.[21] The spatial isolation created by this kind of confinement affords controls on the animals in terms of what they eat, and protects them from certain diseases and predation.[22]

This exertion of control, away from the less regulated environment of the farmyard and toward the more highly regulated one of the broiler house, operates as a necessary component for create the chicken of tomorrow if it is to possess the desired qualities. Allowing chicken access to the outside poses risks, both to health and to consistency. Obvious risks include predators, both aerial and land-based. But there are also risks associated with contact with the less controlled environment and other animals, including interactions with wild birds and other unregulated disease vectors.[23] Consistency of diet could also be compromised, if omnivorous chickens were allowed to browse on their own, which could limit the effectiveness and efficiencies of an engineered diet, and create muscle protein that is less desirable because it has been overly exercised. In this logic, chickens should not feed themselves; instead, they should be fed, and confinement to these spaces allows for that kind of relationship.

It is important to understand that this input-output configuration is not simply a question of mechanics—of getting feed to the birds and removing waste. It is also connected to the wider environmental context, where this machine circulates within those processes external to its workings and needs to be sustained. Agricultural elements become implicated, as feed has to be grown and processed to become an input. Waste as an output does not just have to be collected; it also must be disposed of. At the scales associated with this vision of tomorrow, the energy involved to facilitate these transactions increases considerably. Feed does not just come in, with waste going out. Rather, feed is brought in and distributed, and waste is carried out and deposited. It is important that we notice this translation of energy from individual chickens (who would otherwise browse for their own food and spread their own waste)[24] to systems designed and operated by humans. These systems even had to operate as a partial replacement for the sun. Birds kept in covered enclosures lacked exposure to sunlight and could develop vitamin deficiencies, for example. But with the preference for the controls afforded by this kind of confinement, growers and producers solve the vitamin problem by supplementing feed.[25]

The environment for the chicken of tomorrow does have a futuristic cast, as well. It involves more automation, enabling scheduled watering and feeding times, and further routinization of the lives of the birds. This automation is not automatic, however, as these routines have to be scheduled, organized, and arranged by humans on behalf of the newly designated broilers. Thus, the intertwined protein machines of the Chicken of Tomorrow contest linked the selective breeding of the chicken with the development of a more highly-controlled and standardized environment that surrounded these animals—technologies and techniques of feed and of growth and of processing that maximized the efficiencies of protein production. These environmental transformations also enrolled a larger number of humans into the protein machine, through the newly crafted need for maintaining the means by which the machine functioned—we will explore this component connected to human relationships in greater detail below.

(Component: An Economy)

Another aspect of the machine involved the integration of an economic system that sought to lower the cost of production such that chicken eventually became the cheapest protein per pound. In equivalent dollars, from US$6 per pound in 1948, to US$1.40 per pound in 2014.[26] Controls exerted on the body of the chicken, through the standardization of breeding, environment, and feed, also came to be exerted on the labor of the human grower, through mechanisms of economic incentive and contractual obligation. The human part of the protein machine, that is, also got caught up in its trajectory toward more efficiency and control. Chicken production came to be dominated by so-called integrators—firms that owned all parts of the process, from the hatchery to the feed to the processing plant.[27] Growers became contractors who cared for the firm's property for a short time—taking chicks and growing them according to strict specifications for nutrition and environment. They became facilitators of processes and requirements worked out in advance by powerful producers like Tyson and Perdue who controlled access to breeds, feed, and markets, made demands in terms of productivity, and penalized growers for failing to meet industry expectations. This economic relationship is congruent with the expectations put in place with regard to the animal body and the environment—one that emphasized standardization and consistency to ensure a recognizable final product. Deviation in grower practice, or infrastructure, or feed, might lead to variation in the shape of the chicken or in its timeliness or in its readiness for processing. Firms who integrated and could take more control over every step, then, imposed restrictions on growers in the economic terms of the relationship. This economic consideration also reminds us that humans play a role in this machine, and that the roles are not the same.

(Component: A Set of Human Relationships)

While we have analyzed the intentions that inform the modifications of the chicken and its environment, we have not made explicit the human element in full. That is, humans are involved in setting these parameters and tuning the machine in terms of output. It is important to acknowledge those choices as such and the formation of those intentions. It's also important to note that humans are not just on the outside of this machine designing and tinkering, but are also integral parts of its internal workings as a system, as necessary and irreplaceable components. These roles include

the growers and producers explored above in the newly integrated economic system, but also those who produce and engineer the feed, for
example. In fact, unpacking all of the human relationships involved in the
running of this protein machine adds immeasurable complexity, when
reckoning design, production, infrastructure, processing, distribution,
sales, and consumption.[28] I do want to focus on two key elements here,
however, to illustrate how human relationships are transformed, and help
to shape, the configuration of this emergent protein machine, related to
processing and consumption.

Processing chicken involved a kind of transformation, from animal to
meat—something alive to something consumable. Part of the goals of
standardization impacting the body of the chicken had processing in mind.
Being able to set up automation required birds within a narrow range of
sizes in order to ensure a smooth and consistent workflow. But animal
bodies are never perfectly consistent and even small amounts of variation
have to be managed. In the process of transforming animal into meat,
humans are still required to help deal with the remaining variability (even
as chicken bodies, being small, facilitate greater automation[29]). Thus,
humans still played an integral part of the kill line, for example, hanging
birds upside down and slitting throats, even while other parts could be
automated, like removing feathers.[30] People formed a part of the apparatus, and are themselves altered by the demands of the disassembly line,
having to adjust themselves to the speed and the rhythm of that line.
Importantly, humans can be transformed not just by how the machine
runs, but what outputs emerge from that machine.

Consumption is another part of this equation. The *idea* of a consumer
has been a part of this process from the beginning, operating with a curious kind of duality. On the one hand, producers sought to cater to the
presumed tastes and economic priorities of average American consumers
(by emphasizing more breast meat, for example, and a lower price point).
On the other, producers sought to transform consumers into chicken eaters, through similar mechanisms (emphasizing the less physiologically
complicated breast meat and appealing to the consumer's sense of economy) and by making such consumption less complicated. The former
aspect we have explored above. This latter component comes through an
emphasis on consistency and visibility, achieved in part through branding
and an emphasis on so-called "further processing" (explored below),
which hastens chicken's becoming meat by making it less recognizable as
an animal.

(Component: A Question of Visibility, or the Machine in Context)

Chicken becomes more visible as meat and as a regular part of the American diet, even as the method of producing that protein and the animals critical for doing so slip further from our view. The Chicken of Tomorrow Contest helped to inaugurate a boom in chicken production and consumption among the American populace, a boom that lasted through the next several decades. U.S. per capita chicken consumption has increased every year since 1965, and currently rates at 83.6 pounds per year. Chicken became "meat," and slowly ascended as a popular food, overtaking pork in the 1990s, and supplanting beef as the most consumed meat in 2011. People in the U.S., in fact, eat more chicken now than any other animal protein.[31] So, as a food to consume, chicken has increased in visibility, supplanting other forms of protein on the American dinner table (or drive thru window). How people eat chicken is important—not simply as the recognizable aspect of a body, but also in a variety of processed and branded contexts. Because of production techniques (including breeding, genetics, engineered diets, and structured environments), chicken bodies became more uniform. In part because of shifting settlement patterns (people into cities; animals into factories) and the development of processing techniques (including the creation of processed products like the nugget, the tender, and the patty), the abundant and standardized chicken bodies became less visible as animals.

Like many others growing up in the suburban United States during the 1970s and 80s, I found myself at a strange nexus related to these divergent trends in visibility. My first hourly job as a teenager came as an inaugural employee of a brand-new Kentucky Fried Chicken (KFC) franchise in a suburb of Dallas, where my family had moved more than a decade earlier, away from the farms of mother's youth and toward relative economic prosperity. The South of the 1970s offered the promise of abundant jobs and low-cost housing. The house my parents bought, ironically enough, was part of a development being built on converted agricultural land (primarily cotton), the type that gradually disappeared through my time there—now mostly houses and strip malls and schools, as the population of the Dallas/Fort Worth Metroplex expanded. It was at this job that I first learned that a chicken carcass could be divided into five separate types of pieces—leg, thigh, wing, breast, and what KFC called a center breast or keel. Importantly, KFC, along with similar establishments, was helping to promote and to benefit from the transformation of chicken into a more everyday meat, something consumed on a regular, not simply special, basis.

Thus, the central paradox—as people were becoming more familiar with chicken protein as a meal, we were becoming less familiar with the chicken as an animal. Many families, like my own, were moving away from rural areas or farmsteads. But then, in a sense, so were chickens. Like the larger brooding houses touted by the Arkansas leaflet, the infrastructure for chicken growing was getting bigger and more industrial and more automated. This industrialized environment where the lives of the birds are extensively managed in order to achieve maximum yields and maximum control, helped lead to densely populated areas of chicken farming, at a remove from human population centers. As a result, Ogle notes that the entire process of making meat becomes less visible.[32] As chickens themselves "migrated" from the farm to the factory, they became denizens of a specialized environment, set apart from most humans.

This sense of remove from chicken as animal came not only during the animal's life, but increasingly at its death. Having chicken become meat was a goal of producers—to get consumers to accept chicken as something akin to beef and pork. Becoming meat is also a symbolic transformation, something that occurs to transition an animal into something edible. The process by which animals are killed in a slaughterhouse and become meat, for example, is important. From an event that had historically occurred in proximity to consumers, the processes by which animals become meat are hidden away in an abattoir—a space Vialles has described as housing "an invisible, exiled, almost clandestine activity."[33] As Horowitz has shown, an historical shift has occurred, moving away from live animals being shipped to areas of consumption to be processed by local slaughterhouses and butchers. Instead, grocers began receiving packages of cuts of meat which had been processed at a centralized plant, and in turn would offer smaller prepackaged cuts to consumers.[34] Refrigerated rail cars enabled producers to minimize worries about spoilage, and process animals at a geographic remove. This migration toward a centralized, factory-based efficiency helped to increase the distance—spatially and conceptually—between most humans and meat animals.

For chicken, as well as other forms of meat, such a transformation away from animal is heightened through what is called further processing—transforming the chicken carcass into a variety of pre-cut pieces and recombined forms. By 1996, for example, 86% of chicken sales were cut-up and value-added products[35]—another way these creatures cease to be animals and become meat. For Schatzker, this trend is the consequence of a lack of flavor—as chickens become younger, fatter, more engineered birds, devoid of much flavor themselves, such taste must be imported into the product, through flavorings, breading, and additives.[36] Horowitz has argued that such processing helps to establish branding and product identification.[37] Whatever the reason, meat from chicken could be recombined into a variety of shapes and sizes. For example, the McNugget was introduced in 1983, and at KFC, the Chicken Little sandwich made an appearance during my tenure as an employee, in 1987. And there remains the confused nomenclature of the chicken finger. Like other meat protein, chicken becomes increasingly abstracted from its animal origins for consumers—as Striffler phrases it, from a chicken in every pot, to "a chicken nugget in every fryer."[38]

The dual sense of distance (space and form) reinforces what Bulliet[39] has christened the post-domestic era of human-animal relations. This era needs to be distinguished from what he designates as the pre-domestic (a kind of hunter-gatherer existence where animals were an important part of the world, but not an everyday presence) and the domestic (focused here primarily on farm animals which became an integral part of human existence, a way through which life was sustained and experienced). In the post-domestic era, Bulliet argues that, particularly in industrialized countries like the United States and Japan, humans have developed a much more abstracted set of relationships to animals, involving extremes in both alienation (in terms of those we eat) and intimacy (as pets, for example, increasingly become full members of one's family).[40] Fiddes notes, for example, that people increasingly find discomfort when being reminded of meat's animal origins.[41] This sense of unease reminds us that thinking of meat independent of animals is an option—one made possible by the physical and conceptual distance between most humans and protein machines (like so-called food animals). Bulliet's schema works best as a heuristic, a way of parsing different, co-extant means of relating to non-human animals rather than a depiction of historically distinct and exclusive epochs. But Bulliet does help to show—within the duality of distance and intimacy in the post-domestic era—how we can become increasingly separate from, and closer to, animals.

The Chicken of Tomorrow contest, then, should be seen not just as a means of identifying a better breed of meat-type chicken, but also of establishing a means to *create* that better chicken. Selective breeding is just one mechanism that is part of the process. If we recognize the system, however, that is necessary to produce these chickens—if we operate under a more expansive notion of what counts as a protein machine—then we can also include the environmental changes that are required (a greater emphasis on confinement), the dietary restrictions that need to be imposed (limiting the food to which chickens have access, with a focus on food grown and engineered rather than found), and the alterations of practice and economy asked of the growers (becoming contractors to large producers). The experimental controls of the contest became the standards of chicken production more generally, dictated by industry giants, leading to a highly scientized and technologized process of producing protein. This transformation of production into a more industrial form provided a specific vision of the "tomorrow" that would also be generated. No longer a world of inconsistencies and inefficiencies associated with the production and consumption of chicken, but rather one that was filled with consumers who regularly eat the kind of chicken this protein machine provides—cheat, predictable, even somewhat bland[42] meat. But these changes are just a part of the story. Typical of technological designs, as the visions for the future became real, other less anticipated changes occurred that were not planned or desired.[43] Attempts to solve the problems of their present and thus herald a new kind of future produced new kinds of problems. Thus, the challenges of the past are eliminated as much as they are transformed—the problems of inconsistency, freedom, and cost become the problems of consistency, confinement, and cheapness.

The distance that emerges between humans and animals as meat production moves behind the walls of the factory (which itself moves to the peripheries of human settlement) help to keep these emerging issues hidden for a while, but not forever. The chicken of tomorrow became the chicken of today by transforming what chickens are, how they are raised, how they are processed, and how they are consumed. These transformations include reorganizations of space and scale, along with patterns of work and economic relationships, and result in fewer people involved in the production of more and cheaper chicken protein along with many more people involved in the consumption of that protein. This post-domestic reality, as Bulliet terms it, may cause people to be surprised by the connection between animal and meat, as well as what it takes to get that meat to the plate. More to the point, the future that we now inhabit is one where most food is cheap and consistent, and our concerns are less due to privation and more due to abundance.

The success of the industrial system, in other words, helped to create the conditions under which people subsequently expressed concerns about that system. Namely, in no longer being connected to food production, nor worried about the possible absence of food, people who benefited from industrial largess now had the opportunity to question its very assumptions. Ogle, in her 2013 book *In Meat We Trust*, puts a positive spin on this state of affairs:

> ...[W]e reaped the benefits [...] because factory farming freed so many of our parents, grandparents, and great grandparents, and thus us, from the need to grow and process food; freed them, and us, to instead dream big, think deep, and yes, launch crusades. Factory farming's biggest crop is intellectual capital. So, thanks Big Ag [...] for giving us the cheap food that has nourished an extraordinary abundance of creative energy. Now let's do something with, let's decide what kind of society we want [...].[44]

What Ogle captures here, importantly, is the connection between the material realities made possible by industrial agriculture and the resulting potential to use that time to "launch crusades" against that establishment. We might not turn to look at the workings of the system, but what happens when we do?

...WITH FLAWS

Though the larger protein machine's functions became largely invisible through the new spatial arrangements of humans, animals, and processes, when they were even partially revealed, many people did not like what they saw. For all of the successes associated with the creation of this new tomorrow, through the transformation of the protein machine (in both senses) into an efficient engine of productivity, unintended problems also emerged as a result of this new configuration. The flaws connect to all parts of the machine we explored above that were elements of the assemblage—the animal, the environment, the economy, and the sets of human relationships. The number of those problems associated with the production and consumption of cheaply available meat protein are varied and widely publicized. A litany of studies has explored these issues, revealing what had been hidden by convenience or by design. People have documented the historical development of the more industrialized meat production system, as well as the emergence of failures associated with this industrial turn for animals, farmers/growers, farm and processing plant workers, nearby communities and environments, and consumers.[45] One example comes through the development of Concentrated Animal Feed

Operations (CAFOs), which promote gains in efficiency through, as the label suggests, concentration and confinement. CAFOs achieve an economy of scale—producing more meat, more quickly, with fewer inputs and fewer people. These efficiencies have come through exerting greater control over the animal body at all phases of the lifecycle and over the labor of the people associated with meat production, resulting in greater stresses at all points in the process.[46] For example, animals reach desired weight more quickly, in part through the production and administration of specialized diets and antibiotics. This in turn can strain the skeletal structure of the bird which has added more muscle protein than the frame can bear at a stage of development. And antibiotics, while adept at facilitating growth, can lose their effectiveness to fight off disease in both animals and humans. Mechanization allows for greater concentration of production and processing. Existing rural environmental systems, however, can struggle to contain and absorb the waste generated by animal populations rivaling human concentrations in urban areas. And processing plants can operate at such unrelenting pace that human workers develop repetitive stress injuries. The "success" of this system also has costs on the consumer end, related to the wide availability of food that is cheap to buy and manages to hide the true costs of production. The proliferation of cheap, available protein of all kinds has made diets in the United States more meat-rich, leading to increases in heart disease, for example.[47] All of these challenges emerge in the very practices that defined success in the newly configured protein machines.

Within my own chicken-structured analytical lens, I think it is useful to attempt to organize these emergent flaws in the context of a mnemonic device—one that, in the current context, symbolizes a kind of chicken-adjacent term: PEEEEP. These problems are complex, interrelated, multivariate, and seemingly intractable:

- Physiological—the animal body resists uniformity, grows undesirable elements that are non-edible, and exhibits unprofitable behaviors. As Roger Horowitz puts it, industry strives to "overcome the twin obstacles of natural growth rates and variations in size".[48]
- Environmental—CAFOs generate waste and pollute local communities and ecosystems, which is part of what Weis calls the "ecological hoofprint."[49] Globally, they contribute greenhouse gases to the atmosphere.

- Economic—persistent inefficiencies in raising animals for meat, along with the vertical integration of meat production systems, leaving many farmers with much of the financial risk, and little of the reward, for growing animals.[50]
- Epidemiological—health and disease issues plague animals housed in CAFOs;[51] overconsumption of animal products can create long-term negative health effects in humans;[52]
- Ethical & Political—concerns over animal welfare lead to political action.[53]

These problems are not mutually exclusive, of course, and some solutions to particular problem areas (e.g., antibiotics given to animals sickened by confined environments and simple diets) lead to exacerbated problems in other areas (e.g., environmental overuse of antibiotics and resistant bacteria). It's important to realize, however, that these problems emerge with, or are intensified by, the specific kind of industrialized protein machine that became real and dominant in the twentieth century. And the problems are related to how that machine works—how protein is produced and how it is consumed. It is also important to note that, because of this form of organization and efficiency, people have become mostly consumers that rely on the predictable and efficient production of food for which just a few others are responsible.

And yet, people are still being urged to turn and look at what makes contemporary life possible. Such awareness can be deliberate and systematic, relating to a desire to know more about where our food comes from, or it can be accidental. Michael Pollan is perhaps one of the more known contemporary promoters of food awareness, though works like Eric Schlosser's *Fast Food Nation*[54] and writers like Wendell Berry have long urged readers to remember that "eating is an agricultural act".[55] Why certain people choose to look, while others do not, is difficult to parse, though can be influenced (but not determined) by an individual's relative material well-being. It happened to me because of the experience travelling I explored in the introduction. The visibility of this portion of animal production, transporting birds to market, represented my first real confrontation with the scale of animal production, and recalibrated my sense of distance from it. I, like many others before and since, more fully confronted the realities of what it means to be a meat eater.

Growing awareness and concern about these problems associated with making meat has led to a variety of responses, especially engaging the ethical and political aspects of participation in food systems, and which challenge the current configuration of the protein machine. For consumers—a diet based on abstinence from meat, like vegetarianism or veganism,[56] greater awareness of the contexts of production and subsequent altered behavior meant to challenge it,[57] and advocacy for a change in laws governing meat animals.[58] Dietary choices do help insulate the individual from some aspects of the industrialized system, but outside of radical transformation of diets away from meat in a large scale,[59] the industrial system seeks to meet an expanding world-wide demand for meat. Such concerns have led to laws such as in the European Union, where the Farm Animal Welfare Council adopted the Five Freedoms.[60] For producers responding to market pressures or asserting ideals—minor modifications, such as those demanded by McDonald's restaurants from suppliers,[61] or more fundamental changes to animal husbandry.[62] This latter tack involves taking the animal body as the limit of technological intervention, to which the system itself should be beholden. One example is instructively christened grass-farming by a host of individuals[63]—grass being the primary focus of intervention by the farmer. This method unwinds the complexities of the farm/factory system, suggesting that the solution could be toward less intensive farming. Animals are rotated through a grazing schedule, participating in a different environment than the one of the factory, imposing less stress on the animal body. More field, less factory. None are perfect, however, or solve all of the problems. Abstinence offers an individual respite from participation in the system, but leaves the existing structures relatively unchallenged.[64] Practices like grass farming presently occupy a more boutique market niche represented by higher consumer prices, thus complementing rather than challenging the existing structure, are not fully sustainable and still involve the killing of animals.[65] From within the protein machine, this kind of awareness, precipitated by the distance built into its structure, is itself a kind of flaw. Consumers are required for the machine to work, after all, and either must be segregated from the workings or allowed to feel comfortable with them.

Within the logic of the protein machine built, in part, by the Chicken of Tomorrow efforts, chicken becomes visible primarily as a protein machine itself. The focus of the larger system is attuned to producing chicken protein, and all other components of it are dedicated to that singular goal. As a result, those elements seen as nonessential or even detrimental to the final product are exactly the source of the current problems. Animals that want to do more than eat food and put on weight—to browse and forage, to interact socially. Environments that are more than just waste pits or nutrient extraction points—housing communities and supporting life. Humans who are more than repetitive motion machine or a disadvantaged economic class—beings concerned with ethics and social relationships. So, while the protein machine has been successful at narrowing the world to protein production, it has done so by ignoring so many other elements that never went away.

The meat protein industry has helped to create a well-fed populace with a growing concern not for the whether of food, but the what and the how. Meat production had largely disappeared from view—not just the practices of the abattoir and the presentation of the meat, but the entire life-cycle of the animal. These practices were never invisible—the production processes required people, though ever fewer in number, and required space, though further removed from centers of human settlement. But such visibility has grown as people have chosen to be more particular about the sources of their sustenance. The growing awareness of the problems associated with industrial agriculture comes with its own set of complexities—this is an awareness that is made possible, in part, by the self-same system now viewed skeptically or with disdain. One that exerted some measure of control over the natural world in an effort to provide a predictable and abundant supply of food. And one that created and then fed a nation of chicken eaters. So, what we may need to do is recalibrate in this new present, and look to a new chicken of tomorrow.

NOTES

1. Horowitz, *Putting Meat on the American Table*, 117.
2. See Sachs, Carolyn E. 1996. *Gendered Fields Rural Women, Agriculture, And Environment*. New York: Taylor and Francis.
3. Lawler, Adam. 2014. *Why Did the Chicken Cross the World?: The Epic Saga of the Bird That Powers Civilization*. New York: Atria Books: 206. Perhaps not surprisingly, as Adams has explored, with the industrialization of poultry production, as it changed from a chore to a business that could bear economic investment, farm women lost significant control over a major source of their income. See Adams, Jane. 2002. "'Modernity' and U.S. Farm Women's Poultry Operations: Farm Women Nourish the Industrializing Cities 1880-1940." http://www.siu.edu/~jadams/chickens_modernity/adams-pg_1.html. Accessed 17 March 2009.
4. See Striffler, *Chicken*, 214.

5. Barclay, Eliza. "A Nation of Meat Eaters: See How It All Adds Up." NPR.org, June 27, 2012. https://www.npr.org/sections/thesalt/2012/06/27/155527365/visualizing-a-nation-of-meat-eaters. Accessed May 30, 2018.
6. See Schatzker, Mark. 2016. *Dorito Effect – the Surprising New Truth about Food and Flavor.* New York: Simon & Schuster.
7. Dressed here has something of the opposite connotation one might expect, referring to the chicken carcass denuded, processed, and reading for cooking.
8. Audio Productions, *The Chicken of Tomorrow.* See checklist at the 2:24 mark.
9. Barclay, "A Nation of Meat Eaters."
10. See Horowitz, "Making the Chicken of Tomorrow"; Stull, Donald D, and Michael J Broadway. 2013. *Slaughterhouse Blues: The Meat and Poultry Industry in North America.* Belmont, CA: Wadworth Cengage Learning; Schatzker, *Dorito Effect.*
11. Ogle, Maureen. 2013a. *In Meat We Trust: An Unexpected History of Carnivore America.* New York: Houghton Mifflin Harcourt.
12. "The problem of nature… [in part] a set of obstacles to be overcome via the industrialization of avian biology": Boyd, "Making Meat", 664.
13. See Bugos, Glenn. 1992. "Intellectual Property Protection in the American Chicken Breeding Industry." *Business History Review* 66: 127–68; Boyd, "Making Meat."
14. Schatzker, *The Dorito Effect,* 24.
15. Ogle, Maureen. 2013b. "Riots, Rage, and Resistance: A Brief History of How Antibiotics Arrived on the Farm." Scientific American Blog Network. September 2013. https://blogs.scientificamerican.com/guest-blog/riots-rage-and-resistance-a-brief-history-of-how-antibiotics-arrived-on-the-farm/. Accessed 22 May 2018.
16. Harrison, Ruth. 2013. *Animal Machines: The New Factory Farming Industry.* Edited by M. S. Dawkins. Reissued and Updated ed. edition. Boston, MA: CABI: 37–8.
17. Harrison, *Animal Machines,* 38.
18. For example, the modern egg layer, a variant of the Leghorn, became the favored industrial breed in part because she could be easily dissuaded from her urge to brood, and thus could be kept in her production cycle—laying eggs productively instead of just sitting on them (see Smith, Page, & Charles Daniel. *The Chicken Book.* Boston: Little, Brown, and Company, 1975, 239).
19. Horowitz, *Putting Meat on the American Table,* 117.
20. Site of the 1951 Chicken of Tomorrow Contest.

21. USDA National Agricultural Statistics Service. "Agricultural Resource Management Survey: Broiler Highlights," ND. https://www.nass.usda. gov/Surveys/Guide_to_NASS_Surveys/Ag_Resource_Management/ ARMS_Broiler_Factsheet/Poultry%20Results%20-%20Fact%20Sheet.pdf. Accessed 22 May 2018.

22. As anyone who has raised "free-range" chickens, like myself, can attest, the site of one's flock can suddenly and unexpectedly diminish from one day to the next. Some make other choices. See Shannon. 2012. "Free Ranging Didn't Work for Me...". My Pet Chicken Blog. July 11, 2012. https:// blog.mypetchicken.com/2012/07/11/free-ranging-just-didnt-work-for-me/. Accessed 22 May 2018.

23. This sense of regulation and control became a selling point for growers who used enclosed facilities during the avian flu outbreak of 2005. The argument was that these confinements were better than free-range, because the barriers protected chickens (and by extension humans) and helped to slow the spread of the disease. See Tyson Foods. 2005. "American Chicken and Avian Influenza." Tyson Foods, Inc. 2005. http://www.tyson.com/ Corporate/PressRoom/ViewArticle.aspx?id¼1938.   Accessed   22 May 2018.

24. Not totally, of course. In domesticated settings, chickens are still fed. I provide supplemental feed, though it often goes ignored in the summer months. And the distribution of waste is not total, either. But confinement structures move the onus completely over to human systems.

25. See Stull & Broadway. *Slaughterhouse Blues.*

26. See Schatzker, *The Dorito Effect.*

27. See Horowitz, "Making the Chicken of Tomorrow."

28. As far back as 1690, John Locke reckoned up all of the value found in a loaf of bread (value, for Locke, being an expression of human labor): "[…] it is not barely the plough-man's pains, the reaper's and thresher's toil, and the baker's sweat, is to be counted into the bread we eat; the labour of those who broke the oxen, who digged [sic] and wrought the iron and stones, who felled and framed the timber employed about the plough, mill, oven, or any other utensils, which are a vast number, requisite to this corn, from its being feed to be sown to its being made bread, must all be charged on the account of labour, and received as an effect of that: nature and the earth furnished only the almost worthless materials, as in themselves. It would be a strange catalogue of things, that industry provided and made use of, about every loaf of bread, before it came to our use, if we could trace them; iron, wood, leather, bark, timber, stone, bricks, coals, lime, cloth, dying drugs, pitch, tar, masts, ropes, and all the materials made use of in the ship, that brought any of the commodities made use of by any of the workmen, to any part of the work; all which it would be almost impos-

sible, at least too long, to reckon up." Locke, John. 1690. "Second Treatise of Civil Government." https://www.marxists.org/reference/subject/politics/locke/ch05.htm. Accessed 30 May 2018.

29. Horowitz, *Putting Meat on the American Table*, 121.

30. As a curious historical side-note, the breed of chicken that became the industry standard was not the red-feathered Cornish-New Hampshire crossbreed, but the White Plymouth Rock, a two-time runner up—its white feathers were seen as more desirable because they were less visible if a few pin feathers remained after processing (Stull & Broadway, *Slaughterhouse Blues*, 40).

31. One telling statistic—according to Hart, the yearly total of chicken consumption in the U.S. in 1934 (67 million lbs.) amounts to two days' worth presently—see Hart, John Fraser. 2003. *The Changing Scale of American Agriculture*. University of Virginia Press, 112.

32. Ogle, *In Meat We Trust*. See also Novak, Joel. 2012. "Discipline and Distancing: Confined Pigs in the Factory Farm Gulag." In *Animals and the Human Imagination: A Companion to Animal Studies*, edited by Aaron Gross and Anne Vallely, 121–51. New York: Columbia University Press, regarding pigs in Canada.

33. Vialles, Noélie. 1994. *Animal to Edible*. Cambridge, UK: Cambridge University Press: 5.

34. Yates-Doerr, Emily, and Annemarie Mol. 2012. "Cuts of Meat: Disentangling Western Natures-Cultures." *The Cambridge Journal of Anthropology* 30 (2): 48–64. https://doi.org/10.3167/ca.2012.300204.—packaging might highlight where geographically the meat was produced or shipped from, but "do[es] not insist on the lives, times, and killing of the animal" (51).

35. Horowitz, "Making the Chicken of Tomorrow," 227.

36. Schatzker, *The Dorito Effect*.

37. Horowitz, "Making the Chicken of Tomorrow."

38. Striffler, *Chicken*, 16–7.

39. See Bulliet, Richard W. 2005. *Hunters, Herders, and Hamburgers: The Past and Future of Human-Animal Relationships*. New York: Columbia University Press.

40. See Herzog, Hal. 2010. *Some We Love, Some We Hate, Some We Eat: Why It's So Hard to Think Straight About Animals*. New York: Harper.

41. Fiddes, Nick. 1991. *Meat, a Natural Symbol*. London: Routledge: 95.

42. See Schatzker, *The Dorito Effect*.

43. As Paul Virilio notes, "Every technology produces, provokes, programs a specific accident." Quoted in Loeb, Zachary. "Inventing the Shipwreck." *Real Life*. January 3, 2022. https://reallifemag.com/inventing-the-shipwreck/.

44. See Ogle, *In Meat We Trust*, 267.
45. See, for example, with the production system, Horowitz, *Putting Meat on the American Table*; the industrial turn for animals, Weis, Tony. 2013. *The Ecological Hoofprint: The Global Burden of Industrial Livestock*. London: Zed Books; farmers/growers, Philpott, Tom. 2010. "Squeezed to the Last Drop: The Loss of Family Farms." In *The CAFO Reader: The Tragedy of Industrial Animal Factories*, edited by Daniel Imhoff, 176–81. Berkeley, CA: University of California Press, and Novak, "Discipline and Distancing"; workers, Striffler, *Chicken*, and Cook, Christopher D. 2010. "Sliced and Diced: The Labor You Eat." In *The CAFO Reader: The Tragedy of Industrial Animal Factories*, edited by Daniel Imhoff, 232–39. Berkeley, CA: University of California Press; communities, Stull & Broadway, *Slaughterhouse Blues*; environments, Weis, *The Ecological Hoofprint*; consumers, Lustgarden, Steve, and Debra Holton. n.d. "What About Chicken?" EarthSave.org. Accessed May 24, 2018. http://www.earthsave.org/health/what_about_chicken.htm, Accessed 24 May 2018, Ogle, *In Meat We Trust*, and Schatzker, *The Dorito Effect*.
46. See Imhoff, Dan. 2010. *The CAFO Reader: The Tragedy of Industrial Animal Factories*. Healdsburg, CA: Watershed Media.
47. Importantly, chicken has actually benefited from these concerns, touted as a leaner and healthier form of meat protein—so much so that pork producers started running ads in the 1980s proclaiming pork to be "the other white meat." Of course, in its further processed version, chicken can lose some of its claim to being a healthier alternative.
48. Horowitz, *Putting Meat on the American Table*, 131.
49. Weis, *The Ecological Hoofprint*.
50. The chicken industry pioneered this integration—see Striffler, *Chicken*; it has spread to other meat production industries—see Leonard, Christopher. *The Meat Racket: The Secret Takeover of America's Food Business*. New York: Simon & Schuster, 2014.
51. Kirby, David. 2011. *Animal Factory: The Looming Threat of Industrial Pig, Dairy, and Poultry Farms to Humans and the Environment*. New York: St. Martin's Griffin.
52. Simon, David Robinson. 2013. *Meatonomics: How the Rigged Economics of Meat and Dairy Make You Consume Too Much–and How to Eat Better, Live Longer, and Spend Smarter*. San Francisco, CA: Conari Press.
53. Joy, Melanie. 2010. *Why We Love Dogs, Eat Pigs, and Wear Cows: An Introduction to Carnism: The Belief System That Enables Us to Eat Some Animals and Not Others*. San Francisco: Conari Press.
54. Schlosser, Eric. 2012. *Fast Food Nation: The Dark Side of the All-American Meal*. Revised ed. edition. Boston: Mariner Books.

55. Berry, Wendell. 2009. "The Pleasures of Eating." Ecoliteracy.org. Accessed May 23, 2018. https://www.ecoliteracy.org/article/wendell-berry-pleasures-eating.
56. See Foer, Jonathan Safran. 2009. *Eating Animals*. New York: Little, Brown, and Company; Pluhar, Evelyn B. 2010. "Meat and Morality: Alternatives to Factory Farming." *Journal of Agricultural and Environmental Ethics* 23 (5): 455–68.
57. Keith, Lierre. 2009. *The Vegetarian Myth: Food, Justice, and Sustainability*. 1 edition. Crescent City, CA: PM Press.
58. Tomaselli, Paige, and Meredith Niles. 2010. "Changing the Law: The Road to Reform." In *The CAFO Reader: The Tragedy of Industrial Animal Factories*, edited by Daniel Imhoff, 314–29. Berkeley, CA: University of California Press.
59. A Harris Poll conducted in 2015 found that around 4% of the U.S. adult population ate either a strictly vegetarian or strictly vegan diet. The Vegetarian Resource Group, "How Many Adults in the U.S. Are Vegetarian and Vegan," n.d. Accessed January 14, 2020. https://www.vrg.org/nutshell/Polls/2016_adults_veg.htm.
60. European Commission, "Animal Welfare on the Farm – Food Safety." n.d. Food Safety. Accessed May 23, 2018. https://ec.europa.eu/food/animals/welfare/practice/farm_en.
61. Michel, Melodie. 2012. "McDonald's US Pledges to Phase out Sow Stalls." Globalmeatnews.com. Accessed May 23, 2018. https://www.globalmeatnews.com/Article/2012/02/16/McDonald-s-US-pledges-to-phase-out-sow-stalls.
62. Logsdon, Gene. 2004. *All Flesh Is Grass: The Pleasures and Promises of Pasture Farming*. 1 edition. Athens: Swallow Press.
63. Salatin, Joel. 1996. *Salad Bar Beef*. 1st edition. Swoope, Va: Polyface.
64. The issue of inconsequentialism related to global problems: Vogel, Steven. 2015. *Thinking like a Mall: Environmental Philosophy after the End of Nature*. 1 edition. The MIT Press.
65. McWilliams, James E. 2012. "Opinion | The Myth of Sustainable Meat." *The New York Times*, April 12, 2012, sec. Opinion. https://www.nytimes.com/2012/04/13/opinion/the-myth-of-sustainable-meat.html.

Bibliography

Adams, Jane. "'Modernity' and U.S. Farm Women's Poultry Operations: Farm Women Nourish the Industrializing Cities 1880–1940." Yale University, 2002. http://www.siu.edu/~jadams/chickens_modernity/adams-pg_1.html.
Audio Productions. *Chicken of Tomorrow, The*, 1948. http://archive.org/details/Chickeno1948.

Barclay, Eliza. "A Nation Of Meat Eaters: See How It All Adds Up." NPR.org, June 27, 2012. https://www.npr.org/sections/thesalt/2012/06/27/155527365/visualizing-a-nation-of-meat-eaters.

Berry, Wendell. "The Pleasures of Eating." ecoliteracy.org, June 29, 2009. https://www.ecoliteracy.org/article/wendell-berry-pleasures-eating.

Boyd, William. "Making Meat: Science, Technology, and American Poultry Production." *Technology and Culture* 42, no. 4 (2001): 631–64.

Bugos, Glenn. "Intellectual Property Protection in the American Chicken Breeding Industry." *Business History Review* 66 (1992): 127–68.

Bulliet, Richard W. *Hunters, Herders, and Hamburgers: The Past and Future of Human-Animal Relationships.* New York: Columbia University Press, 2005.

Cook, Christopher D. "Sliced and Diced: The Labor You Eat." In *The CAFO Reader: The Tragedy of Industrial Animal Factories*, edited by Daniel Imhoff, 232–39. Berkeley, CA: University of California Press, 2010.

European Commission. "Animal Welfare on the Farm – Food Safety – European Commission." Food Safety, May 23, 2018. https://ec.europa.eu/food/animals/welfare/practice/farm_en.

Fiddes, Nick. *Meat, a Natural Symbol.* London: Routledge, 1991.

Foer, Jonathan Safran. *Eating Animals.* New York: Little, Brown, and Company, 2009.

Haraway, Donna. *The Companion Species Manifesto: Dogs, People, and Significant Otherness.* Chicago: Prickly Paradigm Press, 2003.

Harrison, Ruth. *Animal Machines: The New Factory Farming Industry.* Edited by M. S. Dawkins. Reissued and Updated ed. edition. Boston, MA: CABI, 2013.

Hart, John Fraser. *The Changing Scale of American Agriculture.* University of Virginia Press, 2003.

Herzog, Hal. *Some We Love, Some We Hate, Some We Eat: Why It's So Hard to Think Straight About Animals.* New York: Harper, 2010.

Horowitz, Roger. "Making the Chicken of Tomorrow: Reworking Poultry as Commodities and as Creatures, 1945–1990." In *Industrializing Organisms: Introducing Evolutionary History*, edited by Susan R. Schrepfer and Philip Scranton, 215–35. New York: Routledge, 2004.

———. *Putting Meat on the American Table: Taste, Technology, Transformation.* Baltimore: Johns Hopkins University Press, 2006.

Imhoff, Daniel, ed. *The CAFO Reader: The Tragedy of Industrial Animal Factories.* 1 edition. Berkeley, Calif: University of California Press, 2010.

Joy, Melanie. *Why We Love Dogs, Eat Pigs, and Wear Cows: An Introduction to Carnism: The Belief System That Enables Us to Eat Some Animals and Not Others.* San Francisco: Conari Press, 2010.

Keith, Lierre. *The Vegetarian Myth: Food, Justice, and Sustainability.* 1 edition. Crescent City, Ca: PM Press, 2009.

Kirby, David. *Animal Factory: The Looming Threat of Industrial Pig, Dairy, and Poultry Farms to Humans and the Environment.* New York: St. Martin's Griffin, 2011.

Lawler, Adam. *Why Did the Chicken Cross the World?: The Epic Saga of the Bird That Powers Civilization.* New York: Atria Books, 2014.

Leonard, Christopher. *The Meat Racket: The Secret Takeover of America's Food Business.* New York: Simon & Schuster, 2014.

Locke, John. "Second Treatise of Civil Government," 1690. https://www.marxists.org/reference/subject/politics/locke/ch05.htm.

Loeb, Zachary. "Inventing the Shipwreck." *Real Life.* January 3, 2022. https://reallifemag.com/inventing-the-shipwreck/.

Logsdon, Gene. *All Flesh Is Grass: The Pleasures and Promises of Pasture Farming.* 1 edition. Athens: Swallow Press, 2004.

Lustgarden, Steve, and Debra Holton. "What About Chicken?" EarthSave.org, May 24, 2018. http://www.earthsave.org/health/what_about_chicken.htm.

McWILLIAMS, JAMES E. "Opinion | The Myth of Sustainable Meat." *The New York Times,* April 12, 2012, sec. Opinion. https://www.nytimes.com/2012/04/13/opinion/the-myth-of-sustainable-meat.html.

Michel, Melodie. "McDonald's US Pledges to Phase out Sow Stalls." globalmeatnews.com, May 23, 2018. https://www.globalmeatnews.com/Article/2012/02/16/McDonald-s-US-pledges-to-phase-out-sow-stalls.

Novak, Joel. "Discipline and Distancing: Confined Pigs in the Factory Farm Gulag." In *Animals and the Human Imagination: A Companion to Animal Studies,* edited by Aaron Gross and Anne Vallely, 121–51. New York: Columbia University Press, 2012.

Ogle, Maureen. *In Meat We Trust: An Unexpected History of Carnivore America.* New York: Houghton Mifflin Harcourt, 2013a.

———. "Riots, Rage, and Resistance: A Brief History of How Antibiotics Arrived on the Farm." Scientific American Blog Network, September 2013b. https://blogs.scientificamerican.com/guest-blog/riots-rage-and-resistance-a-brief-history-of-how-antibiotics-arrived-on-the-farm/.

Philpott, Tom. "Squeezed to the Last Drop: The Loss of Family Farms." In *The CAFO Reader: The Tragedy of Industrial Animal Factories,* edited by Daniel Imhoff, 176–81. Berkeley, CA: University of California Press, 2010.

Pluhar, Evelyn B. "Meat and Morality: Alternatives to Factory Farming." *Journal of Agriculture and Environmental Ethics* 23, no. 5 (2010): 455–68.

Sachs, Carolyn E. *Gendered Fields Rural Women, Agriculture, And Environment.* New York: Taylor and Francis, 1996.

Salatin, Joel. *Salad Bar Beef.* 1st edition. Swoope, Va: Polyface, 1996.

Schatzker, Mark. *Dorito Effect – the Surprising New Truth about Food and Flavor.* New York: Simon & Schuster, 2016.

Schlosser, Eric. *Fast Food Nation: The Dark Side of the All-American Meal.* Revised ed. edition. Boston: Mariner Books, 2012.

Shannon. "Free Ranging Didn't Work for Me…" Blog. My Pet Chicken Blog, July 11, 2012. https://blog.mypetchicken.com/2012/07/11/free-ranging-just-didnt-work-for-me/.

Simon, David Robinson. *Meatonomics: How the Rigged Economics of Meat and Dairy Make You Consume Too Much–and How to Eat Better, Live Longer, and Spend Smarter.* San Francisco, CA: Conari Press, 2013.

Smith, Page, and Charles Daniel. *The Chicken Book.* Boston: Little, Brown, and Company, 1975.

Striffler, Steve. *Chicken: The Dangerous Transformation of America's Favorite Food.* New Haven: Yale University Press, 2005.

Stull, Donald D, and Michael J Broadway. *Slaughterhouse Blues: The Meat and Poultry Industry in North America.* Belmont, CA: Wadworth Cengage Learning, 2013.

Tomaselli, Paige, and Meredith Niles. "Changing the Law: The Road to Reform." In *The CAFO Reader: The Tragedy of Industrial Animal Factories,* edited by Daniel Imhoff, 314–29. Berkeley, CA: University of California Press, 2010.

Tyson Foods. "American Chicken and Avian Influenza." Tyson Foods, Inc., 2005. http://www.tyson.com/Corporate/PressRoom/ViewArticle.aspx?id¼41938.

USDA National Agricultural Statistics Service. "Agricultural Resource Management Survey: Broiler Highlights," May 22, 2018. https://www.nass.usda.gov/Surveys/Guide_to_NASS_Surveys/Ag_Resource_Management/ARMS_Broiler_Factsheet/Poultry%20Results%20-%20Fact%20Sheet.pdf.

Vegetarian Resource Group. "How Many Adults in the U.S. Are Vegetarian and Vegan." Accessed January 14, 2020. https://www.vrg.org/nutshell/Polls/2016_adults_veg.htm.

Vialles, Noélie. *Animal to Edible.* Cambridge, UK: Cambridge University Press, 1994.

Vogel, Steven. *Thinking like a Mall: Environmental Philosophy after the End of Nature.* 1 edition. The MIT Press, 2015.

Weis, Tony. *The Ecological Hoofprint: The Global Burden of Industrial Livestock.* London: Zed Books, 2013.

Yates-Doerr, Emily, and Annemarie Mol. "Cuts of Meat: Disentangling Western Natures-Cultures." *The Cambridge Journal of Anthropology* 30, no. 2 (September 1, 2012): 48–64. https://doi.org/10.3167/ca.2012.300204.

CHAPTER 3

An Interval: A Chicken, Ramona

Abstract This short section tells the story of Romana, a chicken who consistently exposes the flaws in my understanding and the limits of my knowledge. This kind of interaction reveals another way humans and chickens can co-exist.

Keywords Chicken • Backyard • Surprise

My original attempt to raise chickens in Virginia having gone awry, I moved to upstate New York and subsequently abandoned my nascent status of hobbyist poultry producer for a spell. The climate up north was more intimidating, my life was more chaotic, and my living situation was less stable. But after a several years long hiatus, during which I acclimated myself to my new environment (climate, work, and life), I sought to incorporate chickens back into my routine. In particular, after consulting with local chicken hobbyists who managed despite the frigid and snowy winters, I decided to try again. My previous flock was all the same breed (Red-Star Sex-linked), and I admittedly had trouble telling them apart. This time, I ordered six distinct varieties, based first on cold-hardiness and then on looks. They were born in Missouri and shipped, overnight, to the local post office, where they were picked up, just barely a day old. A friend who had ordered with me retrieved them, and I took possession later that afternoon—six chicks in all (a Buckeye, a Buff Brahma, a Buff Orpington, a Gold-Laced Wyandotte, an Australorp,[1] and a Dominique). I set them up in a makeshift brooder in my office, with a heat lamp, food, water, and a bit of room to roam.

A few weeks later, after much chick-like behavior (this amounts to peeping and eating and drinking and pooping and sleeping in what seemed like three-minute cycles), the chicks started testing the boundaries of their confines.[2] To what extent were the chicks actually contained? They would hop on whatever was slightly higher than the ground, limited only by their developing jumping ability and their fearlessness—first the feeder, then the water dispenser, and then eventually, and perhaps inevitably, the edge of the brooder. One evening, while I was working in the office, one chick made this circuit and decided to take it a step further, by leaping onto my lap. What her next move was, I could not say. Too high to leap back down into the brooder, too low to access anywhere else. In fact, I asked her what she intended to do next: "So, what's the plan here?" She had no response. So I returned this little speckled black and white chick—the Dominique—to the rest of her cheeping and somewhat jealous brood.[3] This was Ramona.

She was the first chicken to receive a name. Others soon followed, including Arnold, the Buff Brahma, and Cheddar, the red-feathered Buckeye. But Ramona, three years old at the time of this writing, is still quite the precocious chicken. She has survived the loss of four of the original six (by mysterious disappearance, by illness, by hawk, and by car), and one of the next three (another hawk). In fact, the first hawk took the Buckeye while Ramona and the Australorp took shelter in a nearby collection of brambles, emitting the closes thing in "chicken" to a call for help. Her survival up to this point is all the more remarkable given that she is the one chicken most likely to hunker down, wings spread slightly out, and stamp her feet at the first sign of danger. Which, frankly, makes it very easy to pick her up. She does not seem to mind too much. This I found to be fortuitous, because she has also survived a case of bumblefoot on both feet. Despite the quirky and clumsy name, bumblefoot can develop into a serious, life-threatening infection that makes it painful for a chicken to walk. After noticing her limp and consulting several internet forums, I had a diagnosis and an intervention plan. This consisted of a relatively gruesome procedure that involved a pair of tweezers, alcohol and antiseptic spray, the removal of a scab and then the staph infection, the cleaning of the wound, and finally the application of several dressings' worth of gauze and bright blue vet wrap for her feet. She cut quite the figure in blue spats (see Image 3.1).

Image 3.1   Ramona post foot surgery, image by author

I am somewhat amazed that she is still alive. A lot of hazards out there for chickens, especially (but not limited to) those who have a free-range lifestyle. This freedom is something that it could be said I allow, in spite of the risks—I reflect on the trade-off each time I have a close call or an actual loss, but on balance let the chickens decide how much risk to take. And I have helped keep her alive, to a degree, by remembering to latch the coop at night, noticing a limp, and throwing a few prized blueberries her way (which she will fight you for, and win). Mostly, though, she has survived because of her. Who knows—she might call it luck. I suppose we could ask her.

Notes

1. The Australorp was actually intended to be a Jersey Giant. My friend had ordered the Australorp. But, as chicks, it was difficult to tell them apart—both are black-feathered birds as adults, with no clear distinguishing features in their down as chicks. The biggest difference is, as adults, the Jersey Giant

is just that. Nevertheless, several weeks later my friend called to ask if I thought I had the Giant as intended, because hers was getting a tad large. We never switched back.

2. Much like the velociraptors in "Jurassic Park," though without the malicious intent.

3. I realize that I am anthropomorphizing here, but it is hard not to associate the pitch of the chicks' cries as tinged with jealousy for having made it that far.

CHAPTER 4

A Chicken, Part II

Abstract The second protein machine picks up the story in the last third of the century, when the United States became a nation of chicken eaters, connected to PETA's *in vitro* chicken competition in the early twenty-first century. With a structure that parallels the earlier version of a chicken of tomorrow, this chapter follows the protein machine from the factory into the laboratory, presented as a solution to the problems of industrialization. I argue, however, that this move to the lab actually doubles down on attempts to simplify the chicken and to control its biology and its ecology. In this way it is an example of increased responsibility that humans assume over the natural world.

Keywords *in vitro* meat • Laboratory • Control • Responsibility • Solutions

> *Alternatives are everywhere, though they are often invisible.*
> —David Edgerton, *The Shock of the Old: Technology and Global History since 1900*, Oxford: Oxford University Press, 2007

The mid-twentieth century Chicken of Tomorrow contest certainly did help transform its vision of the future into our present. Chicken is now a

W. Galusky, *Protein Machines, Technology, and the Nature of the Future*, https://doi.org/10.1007/978-3-031-08717-2_4

common, consistent, and cheap form of protein consumed by millions of Americans on a regular basis. It has become so through changes in the physiology of the bird and its accompanying environment, which in turn were accomplished through advances in technology and technique. These overall achievements have come with associated costs, however, as each component—body and system—contained elements that exhibited stress in this new configuration. Birds still sought to be birds in full, despite the exclusive emphasis on protein production; environmental systems strained under the burden of feed production and waste storage. In our post-PEEEEP world, those costs, and people's growing awareness of them, have led to increased tensions related to how the current protein machine is organized, with mounting calls to reform and reconfigure these relationships, to improve chicken once again. Thus, part of what has shifted is the meaning of the term "better" as it relates to chicken. The concept in the middle of last century was defined in terms of efficiency, uniformity, and economy—a better chicken was standardized and cheap. Increasingly in the current century (once "better" chicken became simply chicken), the idea may be redefined in terms of environmental, animal and human health, as well as humaneness and justice. Our task will mirror the one we pursued in Chap. 1, but through the dreams and developments attached to this new kind of protein machine—one that seeks to transform how chicken is produced by bypassing the animal body, while catering to an existing chicken-consuming public. Once again, disaffection with the chicken of our "today" is leading to the desire to create a chicken of tomorrow, which requires a new vision for the chicken and a new vision for the future.

CONFIGURING TOMORROW (THE LABORATORY)

By the early twenty-first century, my entire family had followed my lead and started raising their own chickens as hobbyists. My parents, who moved from the suburbs of my youth to rural New Mexico upon retirement, have a flock of over two dozen layers.[1] More curiously, my brother has a very small number of birds (between one and four at any given time) in suburban Los Angeles.[2] They have become like a number of people who sought to shorten the distance between themselves and the source of their food—an extremely modest way to express some personal intention within the larger food system. Importantly, these choices are made within the existing agricultural production system, with the chickens providing a supplement to the food procured more conventionally—often an expression of privilege rather than necessity, costing more than what might be earned through the selling of eggs to nearby members of the community.[3] However, even these small gestures reflect a preference for greater connection between production and consumption, and some level of unease with what has become the status quo. Importantly, those in the business of feeding Americans have taken notice of this sense of disquiet toward existing food systems.

Tinkering within the existing system of protein production has already begun, in terms of marketing, material relationships, and policy. Fast food restaurant corporations have started the process of influencing the practices in their supply chains—McDonalds, for example, has publicized intentions to source eggs only from cage-free producers for use in its restaurants in the United States and Canada by 2025.[4] These moves come in the context of increased competition from newcomers to fast casual dining like Chipotle, who have made responsible sourcing integral to its entire brand identity.[5] Consumers who possess the requisite economic flexibility can now exercise more choice at the supermarket in terms of what to buy—with more brands offering a variety of options that might appeal to the conscientious shopper.[6] And while policy in the United States at the federal level is limited, places like the European Union have long touted initiatives such as the Five Freedoms, which were formally adopted in 1998 and apply to "the protection of animals kept for farming purposes."[7] These policy initiatives are a means of suggesting that the values that apply to farm animals are not simply a matter for consumers to determine, but for citizens to proclaim.[8]

Another approach to the flaws associated with the current protein machine has sought to bypass expressions of consumer values, or policy interventions, or minor adjustments to the existing factory-based agricultural setting, and instead to move the operation from the factory into the laboratory to solve those problems associated with industrialization—*in vitro* or cultivated meat. Taking advantage of innovations related to stem-cell technology, *in vitro* meat efforts have gained steam in recent years, with the aim of producing meat protein without the need to rely on the animal—to remove this smaller version of the protein machine and reconfigure the larger one. This transformation looks to reconceptualize and reorder how meat is made—not on the skeleton of animals, but in the controlled confines of laboratory facilities; not within a managed ecosystem, but within a sterile container. These efforts, in fact, and the promises they offer for building a better future with better chicken, lead us to the next part of the story.

Opening the playbook from the mid-twentieth century, we can look to a contest as a means of producing better chicken, of finding a new chicken of "tomorrow" and a new tomorrow. The context in which this contest is staged has radically changed, as has our conception of the problem. That is, the issue at hand is not related to the fact that people are *not* eating chicken as much as producers might hope (and thus desire to change the chicken in order to change the consumer), but that people are eating more chicken than one can sustain within the current configuration (and thus desire to change the chicken in order to cater to the existing consumer). These are concerns that have emerged out of the context of an established industrialized form of meat production that feeds a nation of chicken consumers. We might rely on the actions of consumers through market choices (buying more "humane" meat at a higher price point or abstaining from meat altogether), or of citizens through policy initiatives (requiring certain standards of care or outright prohibition), to shift industrial practices. However, there can be limitations to these methods, in terms of their pace, their scale, and their reliability. Another option would be to recreate the protein machine once again, aligned with these new considerations. How might we use a contest to encourage the development of this form of better chicken?

CREATING A PROTEIN MACHINE...

Unlike the previous contest (sponsored initially by A&P Food Stores), this version was not sponsored by a supermarket chain, but by an organization more aligned with this new conception of better—the People for the Ethical Treatment of Animals (PETA). The goal was not to create *a better chicken* in order to grow a small market (how do we make chicken something that consumers will come to want). Rather, the goal was to create *better chicken* in order to feed an existing market (how do we make chicken in a way that satisfies both existing human tastes and developing ethical concerns). To put it another way, what might enable us to have our chicken and eat it too? *In vitro* production technologies. Growing meat *in vitro* moves the production process into the laboratory, relying on the cultivation of undifferentiated muscle cells (myoblasts) within a suitable medium to produce a bounty of pure protein disconnected from the body of the animal. Leveraging techniques refined in work with stem cells, *in vitro* meat is created through harvesting these starter myoblast cells from the desired animal, and then stimulating the growth of those cells to produce protein. By focusing on the muscle sans animal, the aim of this technique is to be able to produce an abundance of protein in a highly controlled environment without having to kill animals to do so.

Presented as a compromise between the reality and intractability of human tastes and the need to reform industrialized methods of meat production, *in vitro* meat technology formed the centerpiece of PETA's 2008 contest. This challenge promoted the technology as a means of enabling the consumption of meat while eliminating the suffering and the environmental compromises typically attached to it, particularly in its industrial form. This move—allowing humans to eat meat without eating animals—has the potential to bypass the animal origins of meat in the creation of protein. Instead, the animal becomes an almost purely abstract component of meat consumption. One might find a whisper of connection to the animal origins of one's meal, if only in the name, but through this technology, there would be no necessary link to the death of an animal. Thus, PETA sought to reward those researchers and investors who could produce a chicken protein that was indistinguishable from more conventionally grown chicken in every way (in taste and in price), save one very important feature: this meat would be grown *in vitro* instead of *in vivo*. Put precisely, official contest rules stated that the winning entry would have to:

1. Produce an *in vitro* chicken-meat product that has a taste and texture indistinguishable from real chicken flesh to non-meat-eaters and meat-eaters alike.
2. Manufacture the approved product in large enough quantities to be sold commercially, and successfully sell it at a competitive price in at least 10 U.S. states.[9]

These are very high benchmarks for success, given the state of the technology at the time (which we will discuss below), focusing on taste, texture, and commercial viability. A winning entry had to be a fully-formed competitor to more conventionally produced chicken (one that relied, that is, on the animal as a platform for protein production). It had to meet existing government requirements for sale, and pass muster not just with a panel of taste-testers, but also U.S. consumers, in a time frame of roughly six years (from time of announcement to closing date). The contest, then, represented an incentive to move the technology passed the stage of promising technique struggling with refinement and efficiencies associated with the process, and to the stage of viable competitor to industrial meat production in the contemporary market place.

How does *in vitro* meat work specifically? Creating muscle protein in this manner involves several steps that would enable researchers to bypass the animal body mostly (if not entirely). In basic outline, the technique involves harvesting myoblast cells from a donor animal, cultivating those cells in a suitable medium (this medium requires the capacity to feed the cells, create an architecture for them to grow in abundance, and stimulate the cells to grow and take on a muscle-like structure), and using the resulting cells to formulate an edible substance (something that resembles the texture of meat). Theoretically, the process is relatively straightforward and there have been incremental successes in producing meat protein in a laboratory setting. The practicalities of creating meat (rather than just an aggregation of muscle cells), however, can lead to a lot of stumbling blocks. Once those starter myoblast cells have been harvested, the remaining steps connected to taste, texture, and commercial viability each pose particular challenges related to satisfying the terms of PETA's contest rules.

Getting cells to grow is not the problem, but getting them to grow in sufficient quantities and with the right kind of texture is. Take perhaps the most important premise—muscle cells can be grown without the need to use animals. Typically (and, upon reflection, that word reads like a major understatement), muscles grow within bodies. To attempt to grow muscles outside of this more usual context, researchers have to create structures that simulate the necessary components to growth that bodies provide. On a typical summer day, my chickens spend it like this: they get up in the morning and whine to be let out of their enclosed run, then spend the rest of their time foraging for food, hiding from any potential threat, and engaging in some type of grooming. They also poop. Through all of that activity, muscle is produced and maintained.

In the factory, this kind of activity is constrained, sometimes radically so, but the basic function of the body and the placement of the muscle remains the same. In the lab, one has to determine which specific elements of these activities are necessary and which are not, and how to best simulate those relevant functions. For *in vitro*, these techniques have to find a way to replace the body—feed the muscle and exercise the muscle outside of its original context. Feeding muscle involves not just nutrients, but a means to get those nutrients to the muscle cells. Nutrient delivery, for example, becomes a problem of engineering. Muscles in bodies have a vascularization system to keep cells fed even in density, but lacking such a system, body-less cells face a spatial challenge. They can either go out (spread thinly on large sheets to prevent inner cells from dying due to lack of nutrients) or go up (attached to an edible scaffold that allows more height for the cell culture). Another set of engineering problems relate to stimulation and striation. Bodies move through space, and provide stimulation to cells through that motion. How does one exercise cells sans body? One option involves flexing those thin sheets to simulate motion. Another option allows for mild electric currents to be applied to the muscle cells. Muscles also do not exist in bodies in isolation—muscle fibers have to be connected to each other and to bones. Muscles co-mingle with connective tissues and fat deposits. All of these elements go into what we have come to know as meat—rather than a simple substance, meat is a complex mixture of substances. In essence, then, laboratory-based methods of producing protein have to replace the body-as-machine with one fabricated in the lab to replace the essential functions otherwise performed by that body, and supply all of the components otherwise present in that body. Such techniques and components have to be identified and refined in an effort to produce meat that is recognizable to humans as such (rather than some kind of gelatinous mass—more on this below).

The technique for re-making meat, by focusing on a more basic cellular component and building up from that (rather than harvesting what is generated from a body), also represents a much more controlled method of production. The burdens or obstacles discussed above can also be configured as a benefit—one may have to build a scaffold or add back other components of meat (e.g., fat or connective tissue), but can do so reflecting specific values or choices—specific kinds of fat, or fat with specific qualities.[10] Nothing need be present that designers do not want to be present—everything results from a decision. Meat has been stripped down to its cellular essence, and built back with what humans have chosen, ideally, to add.

Providing nutrients is not just a structural problem, but also a conceptual one. That is, a large part of the promise of the *in vitro* technique is the removal of the body of the animal in all phases of the production, once the initial harvesting of the cells has taken place. While invasive, this harvesting process can be done without sacrificing the life of the donor animal. However, the current method for growing muscle cells *in vitro* relies on a serum derived from fetal bovine blood, which is harvested from bovine fetuses typically concurrent with the slaughter of the pregnant cow.[11] This serum is a commonly used element for the growth of cells in culture and thus complicates the relationships of animals to the process (not just its development, but also its practice). As such, this is another technical hurdle that will need to be overcome in order to deliver on the promise of animal-less meat. Researchers have faith that non-animal alternatives can be developed,[12] but current research is proceeding using the fetal bovine serum.[13]

The requirements of the contest, with its focus not just on technical, but commercial, viability, also point to a few more major challenges, including texture and price. Texture, for example, is another context-based problem, because the goal of the contest is to produce a product made *in vitro* that is equivalent to muscle protein made more conventionally. The output of cells produced on sheets or attached to scaffolds does not have the same physicality as those within a body. Those embodied muscle cells record motion and the particularities of diet, and this provides the resulting meat with a mouth feel and a flavor that people come to expect. As we explored in the previous chapter, the factory-based protein machine attempted to make those bodies as standard and uniform as possible, though never perfectly. In the lab, however, with no animal body to work with, those cells are not collected in muscle fibers that need to be connected to other muscles or to bones, and are not nestled next to fat deposits. In a certain sense, bodies are mixed media, and muscles intermingle with connective tissue, blood vessels, fat, and bones. These elements help contribute to the texture and flavor of animal protein consumed more conventionally. In the lab, muscle cells are in isolation. Other elements have to be added—part of the promise of this technology is the amount of control that one can exert on the composition of the product. But the means to take a collection of muscle cells and achieve a texture comparable to animal muscle from bodies remain elusive. Those who have consumed *in vitro* produced meat products have compared the texture to jelly and to cake.[14] Meat jelly or meat cake seems like a difficult sell.

Like many early stage technological developments, *in vitro* meat is prohibitively expensive to produce for broad consumption. A proof of concept lab-grown beef burger, for example, cost over US$300,000 to produce. Chicken became more popular as a form of protein in part because the innovations brought the price of the meat down, to the point that it is one of the cheapest forms of animal protein available. For researchers and developers working on *in vitro* varieties, the assumption is that the price will go down as the techniques become refined.[15] Importantly, the terms of the contest suggest that the major goal of the promotion is to produce something that is viable and acceptable to meat eaters. People will need to be willing to buy it and eat it.

Acceptability is an important concept, though one that looks to be settled by innovation and comparability rather than rhetoric. That is, the *in vitro* meat efforts face similar challenges to the chicken producers in the 1940s and beyond—creating a substance that is comparable enough to what people expect (in price, in consistency, in appearance) that they will be willing to incorporate it into their diets. The technological challenges are vast. But people are used to eating food that is disconnected from its origins, and so may be willing to give this kind of meat a try if it resembles what they are used to closely enough. This kind of replacement is central to the PETA contest's requirements—not just producing animal-less meat, but meat that is indistinguishable in taste, texture, price, and availability from its more conventional rivals.

Despite all of these challenges, many of the technology's developers and advocates believe it to solve the myriad problems associated with industrial meat production.[16] A U.S. group called New Harvest promotes *in vitro* meat as solving the following issues:[17]

1. Composition—can control fat content, adding back in only specific fats, in specific quantities, deemed desirable by current dietary standards;
2. Disease control—reduce unsanitary conditions by eliminating waste generated by animals or ecological systems, replaced by cells in media in labs;
3. Efficiency—"Inedible animal structures (bones, respiratory system, digestive system, skin, and the nervous system) need not be grown";
4. Exotic meats (rare and extinct)—the process could be applied to any starter muscle culture, thus allowing for any manner of protein cultivation;
5. Reduction of animal use—cell lines could be cultivated from a single animal.[18]

More to the point, this form of production eliminates several things. It eliminates the uncertainty—the risk—of natural foods. As indicated by a spokesperson for a processing facility, "natural ingredients are a 'wild mixture of substances created by plants and animals for completely non-food purposes—their survival and reproduction.' These dubious substances 'come to be consumed by humans at their own risk.'"[19] It eliminates the environment in which animals would normally live. It eliminates wasteful translation of energy into non-consumable elements. It eliminates moral prohibitions against eating exotic animals, or animals at all, because it eliminates the animal. And, what's more, it eliminates the need for consumers to change. Little else, that is, except a willingness to eat meat from this process.

(Component: A ("Chicken") Body)

In the context of the *in vitro* approach to creating chicken protein, the animal body as it is more commonly understood is not present. As we explored above, this removal is a major part of the impetus for pursuing lab-grown meat, as a means of eliminating the flaws closely coupled to that body—suffering, inefficiencies, waste. This approach is reminiscent of the one taken by the mid-century Chicken of Tomorrow contest—a view of the existing animal body as problematic (inconsistent and inefficient) leads to efforts aimed at creating a new one that does away with those liabilities. For the previous contest, a chicken with a non-standard body with less breast meat that was too expensive to bring to weight was a problem, and was fixed through the cultivation of a new breed of chicken expressing the desired characteristics in terms of physiology, uniformity, and rapid growth, and a support structure to nurture those characteristics. As that body has become a problem in the contemporary world (in part because of the burdens associated with large breasts, quick growing times, and the confinement imposed to help ensure consistency), we might now pursue a new means of generating protein that bypasses those problems. Maybe we can change the body once again, by removing it completely as a part of the protein production process.

It is important to realize, however, that the body of the animal is not simply removed, but rather it is replaced in a more mechanized form. That is, in the *in vitro* context, the animal body with all of its complexities needs to be replaced by a lab-based "body" that does one primary task—support the production of protein. The animal body that does many things (some desirable and others less so) is replaced by a kind of artificial body that has fewer tasks—a function specific alternative. This move represents a kind of ideal—the very goal of previous animal-based agriculture has been to reduce the function of the body to the exclusive production of protein, but the body has been more recalcitrant to such a simplification. Hence the move from animal to lab-based body. Even in this simplified form, however, the goal of a substitute body leads to some profound complexities. Hannah Landecker has explored the emergence of biotechnology and the cultivation of biological matter *in vitro*, in her 2007 book, *Culturing Life*. These are the methods that have enabled the research into lab-based meat, and borrow heavily from the pioneering tissue cultivation techniques. She notes, for example, that the capacity to remove cells from their original contexts and in some ways direct the timing of their production has, in her words, "[changed] what it is to be biological."[20] These types of abilities have made biological matter much more plastic—changeable in the context of human intention, allowing us to dictate how rapidly cells divide, and even when, through the capacity to freeze and then thaw cells to restart their productivity. Importantly, however, these successes in taking and using cell cultures out of the organic bodies in which they were found are only made possible through recreating a new home for those cells. As one lab puts it, "Tissue culture [requires]… the creation of 'a new type of body in which to grow a cell.'"[21] So the body of the organism is replaced by a human-contrived version that needs to replicate the vital components more typically done in the ontological context of the animal—e.g., nourishment, circulation, movement. As we explored earlier in the context of culturing myoblasts to resemble meat, each of these tasks or requirements pose particular challenges as researchers attempt to mimic animal bodies in this reduced form, building up what is critical and omitting what is not (while simultaneously being able to decide which is which).

Aside from structuring and layering protein to simulate muscle, another key role that a body plays is as a boundary to the outside world. The idea of a body can be understood, in part, through its capacity to create a kind of barrier between inside and outside, as a means of maintaining integrity. That boundary can be conceptual,[22] and, critically, it can also be physical. Bodies that function well regulate the ingestion of foreign objects inside, allowing the uptake of nutrients, disposing of waste, and defeating unwanted intruders. Even in the context of a body, of course, these boundaries are not perfect and can lead to breakdown and illness. Part of the challenge of the factory-based protein machine, in fact, is found in the failure of animal bodies to regulate illness well enough to meet the priorities of consistency and efficiency. Stresses connected to confinement and diet led to animals having difficulty fending off disease—even diseases that, pre-confinement, did not pose much of a threat. Smith and Daniel discuss the issue of coccidiosis at it relates to chickens in confinement.[23] Coccidiosis is a common intestinal parasite, and it causes what intestinal parasites are known to cause in all manner of animals, including humans (I presume the reader's personal experiences here will suffice as illustration). For chicken producers looking for maximal and consistent yield, even mildly sick birds cause disruptions and should be avoided, if possible. One solution, in the factory context, involved the use of wire floors that limited the birds' exposure to the parasite found in fecal matter. Prior to this innovation, a chick might be exposed to that parasite early on and develop an immunity to the disease that it carries into adult life. Paradoxically, birds became more prone to this illness as adults due to reduced early exposure, which created even more disruption to the production cycle. And to combat the illness, more medicines were introduced to help the bodies do their regulatory work. The bodies of these birds, however, in being exposed to antibiotics so often and so early (to prevent illness and promote growth) end up as very precarious barriers to disease overall, and lead to the need to impose biosecurity measures in order to limit exposure to external pathogens.[24]

Animal bodies, then, acts as permeable boundaries to the outside world and imperfectly regulate disease. In the lab setting, where the technical body replaces the animal body, this boundary becomes even more tenuous. The current cultivation technique that we explored above involves creating an extremely fecund environment in which to stimulate the growth of these muscle cells. This serum is densely packed with nutrients to encourage the production of new cells in order to generate protein. But such a nutrient-rich environment would be attractive to any biological matter—not just what researchers desire. The problem of contamination by foreign objects does not go away, and there are no internal, body-based systems in place to help root out intruders. Instead, these lab spaces require absolute sterility to maintain their integrity. Importantly, for the viability of this technology, not only is an extreme level of control a virtue, it is also a necessity. The productive environment has to be kept pure, lest unwanted substances take advantage of such a fecund context and corrupt the process. So, for example, because of how fertile these bioreactors are, as one researcher relates to Michael Specter in an article for the New Yorker, "We need completely sterile conditions. If you accidently add a single bacterium to a flask, it will be full in one day."[25] The body as a boundary can also be surpassed through this technique, overcoming the need to constantly produce new animals in order to generate protein. One potential efficiency can be found in the relative permanence of this new, lab-based body. We would no longer need to be bound by an individual body's lifecycle. Previously, the animal body was conceptualized as the machine that produced protein (and itself), but had a clear lifespan that had to be restarted in new bodies. With *in vitro* meat, the lab becomes the body, the machine, which does not have a clear lifespan, and could theoretically perpetuate protein production much longer. The body of the animal, then, is not eliminated but rather replaced by a lab-based substitute, in order to ensure that we grow what we intend—protein that resembles muscle, and nothing else.

(Component: An Environment)

In vitro meat technologies also alter the direct engagements with the more traditional environmental contexts of industrialized meat production—they do not currently rely on the agricultural production of feed grains, or the use of large waste lagoons. There is also no need to move animal bodies through space—in transporting, slaughtering, processing, and disposing of those bodies in the production of meat. Many of the environmental problems associated with industrialized production can be bypassed. We might suggest that the new lab-based context of protein production blurs the distinction between the body and the surrounding environment—the body is not inside a containment system, but is co-extensive with the containment system. But there are still meaningful distinctions to draw between how nutrients are delivered to the cells in culture (a function of a body) and where those nutrients come from (a function of an ecology).

Environments are still implicated within this protein machine. In fact, the same dizzying array of connections can be found in this configuration of the protein machine as could be in the previous one. The fetal bovine serum must be extracted, refined, and delivered to labs, for example. The methods of stimulation employed to simulate bodily movement through space connect this *in vitro* body to systems of electricity generation. The lab context does not exclude environmental systems, but rather transforms them and their involvement in different manners. They get imported in terms of the siting and functioning of the lab, in terms of the manufacture and use of the serum, in terms of the design, production, and use of the relevant equipment. Networks that ensure biosecurity and energy and feed come into play, as well. As with the body, meat production is not removed from an environmental context, but rather has that context transformed.

What is key to remember is that while the environmental context might become less visible in this form of protein machine, it does not disappear or become less important.

(Component: An Economy)

It is not for nothing that one of the requirements of PETA's contest related to economic viability, mandating that entrants be able to generate an *in vitro* produced product and, "successfully sell it at a competitive price in at least 10 U.S. states." This capacity appears many years away, with experimental labs being able to produce small amounts of protein at very sizable costs. As a means of illustration, take the taste-test organized for an *in vitro* produced burger in London in 2013. This event was billed as an effort to demonstrate proof of concept and edibility, though did so with a large price tag.[26]

Importantly, these are start-up costs, which are traditionally very high and tend to come down as the technology and techniques improve and become more efficient. Hopes remain high, for example, that, with a scaled-up production process, that price can be brought down to US$30 per pound in 20 to 30 years.[27] Given the speculative nature of this exercise, as we project into the future, it remains a challenge to see to what extent such cost savings can be realized.[28] Though, were such dreams of animal-less meat to realize their target, we would certainly find a revolutionized protein industry. One of the ambitions, for example, involves developing a "meat machine" similar in size and shape to a bread machine, that would sit on a counter top and produce the desired protein at the programmable push of a button. Slotted into the existing economic system, this protein might go the way of specialized breeds and genetically modified organisms developed by producers, becoming proprietary technology that would need to be licensed in order to be grown.

(Component: A Set of Human Relationships)

Moving production of protein from the factory to the laboratory certainly alters the relationships between people and animals or, more precisely, bioreactors. Lab technicians and experimenters become the primary points of contact with the process, replacing breeders, growers, and processors in the more conventional system. And the knowledge systems that are required to enable the process to work become even more specialized, beyond what Harrison worried about in the previous chapter. The specialists will need to have an intimate knowledge of protein and muscle and tissue cultures. Regimes of expertise will need to be created and utilized. For the untrained and uninitiated, for those disconnected from lab-based protocols, the process of making meat will become even more opaque or invisible than it already is. The lab-setting of production would eliminate the need to have people on the processing line dealing with bodies. Like our contemporary system, though more intensely, the relationships would likely be focused on the consumer side of the experience.

For those consumers who may know little of the process, acceptance will be a key issue. There are groups that seem primed for acceptance: the less finicky (like the ardent fan of the hotdog eating contest), those with more specific needs (like the deep space traveler), or members of the technological or gastronomical avant-garde. However, a large number of people would have to find meat produced in this method as functionally equivalent to more traditional meat production techniques, and some remain very skeptical.[29] Hopkins & Dacey suggest that acceptance should follow from a kind of functional equivalence, arguing that: "What makes meat 'real' is its constituent substance, not its mode of production."[30] This kind of equivalence tends to ignore the larger role that food plays in people's lives, however, and the complexities involved in accepting new foods to eat.[31]

The human relationship to the moral question of meat is also modified. Importantly, the ethical questions surrounding eating meat are not so much engaged as eliminated. People are not asked to confront the ethics of eating meat as they might be[32] in the current configuration—whether in the basic question of killing animals, or in the technologically mediated question of the human, animal, and environmental stresses exacerbated by industrialized systems and capitalist logics. Is there better meat to buy, for example, based on different production or labor practices, reflected on labels and price points? This kind of consumer-based decision making is fraught, especially at the level of trust and accuracy (it is not clear that the labels communicate effectively or can be relied upon to signify what the consumer hopes to support, for example[33]). Rather than confront the complexities and the efficacy of consumer choice or citizen action, *in vitro* meat allows us to bypass these concerns by technological means. The process transforms the challenges a user might face in the context of industrialized modes of production to challenges of design and engineering.

In vitro meat, that is, engages in a kind of ecological modernization, which proposes to solve environmental problems through the design of better technologies rather than the alteration of human behavior. Think electric cars or carbon sequestration programs—technologies that allow us to continue acting the way that we have (driving, flying) with fewer negative impacts because of the design work of others. The protein production technology promoted by the PETA-sponsored contest emphasizes the compatibility or even indistinguishability between its product and more conventional meat, suggesting that consumers shouldn't necessarily notice that they are eating "better" meat (a similar argument is made on behalf of cloned animals[34]). In terms of human relationships, two very different trajectories emerge related to complexity and responsibility. These trajectories are related to an approach to technological design which suggests that the simpler a technology is for the user, the more complex it becomes for engineers and designers.[35] On the one hand, then, users or in this case consumers of meat have their relationships to that product simplified. This comes in terms of the trends we have already seen in the context of consumption related to further processing—we now eat nuggets more often than whole chickens that require more skill in navigating—prepping, cooking, and eating. It also comes in terms of values—questions of suffering and the concerns related to factory farming have been designed away. And any understanding about how precisely meat is made becomes even more obscure to the everyday consumer. On the other hand, researchers and designers have the much more intense burden of managing a complex and precarious system to ensure that meat can be produced in this manner. Simple user, complex system.

(Component: A Question of Visibility, Or the Machine in Context)

Like the intervention with chicken before it, this new technology helps transform the visibility of chicken as meat in various ways. Unlike the contest sponsored by the A&P Foodstores, PETA's version cannot yet be said to inaugurate a boom in this mode of producing chicken. The technology is still in its nascent stages and is relatively unproven, at least at the levels approaching commercial viability. There may be an *in vitro* chicken of tomorrow, but it will not be the winner of this contest, because there was no winner. The deadline has passed for any successful entrant—it does not even appear that there were any entrants at all. It is not difficult to conclude, in fact, that this lack of a winner was by design. Or, rather, the intended winner all along was visibility for PETA's mission to promote animal welfare and to spur interest in this technology as a potential solution to that and other issues. We can draw this conclusion by analyzing several of the criteria required for the winner: commercial viability, the taste test, and the size of the award.

The ability of *in vitro* meat to compete with more conventionally produced protein based on price (part b of the contest rules) is a pretty far-fetched idea. As we explored earlier, even the most optimistic projections still place the cost of *in vitro* meat at over ten times the current price of conventionally produced meat within the next few decades. Of course, a lot could change in the intervening years in terms of pricing structure, including a change to subsidies, breakthroughs in efficiency, and a shift in social values, which may alter the relative comparison in prices. But the timeframe for those projections exist outside of the six-year window of PETA's contest. The second requirement centered on taste (part a of the contest rules), necessitating that entries pass through the gauntlet of a taste test that proved the product to be indistinguishable from conventional chicken, but also that it be cooked using the recipe from PETA's own "fried 'chicken,'" which is a meal preparation developed for a vegetable-based chicken substitute. This aspect of the contest calls attention to the existing recipe and suggests, not too subtly, that a taste-alike version of non-animal "chicken" already exists. Finally, the prize money itself raises red flags—while the US$1 million prize appears relatively robust, it is likely dwarfed by the actual costs required for research and development in the technology. These three aspects—commercial viability in a six-year window, taste-test requirements based on a chicken substitute recipe, and the size of the prize money compared with likely development costs—suggest that the contest was more about promoting the visibility of the problems associated with meat production and the potential of this technology rather than actually crowning a winner.

This conclusion is reinforced by the announcement PETA made after the closing date for the contest. The group posted this update on their website:

> Although the March 4, 2014, deadline for the prize has now expired, PETA's *in vitro* chicken contest was a smashing success! Since we announced the prize, laboratory work on *in vitro* meat has come a long way, and a commercially viable beef hamburger or pork sausage are bound to happen in the not-too-distant future [...]. More good news is that the science used in the

development of *in vitro* pork and beef will eventually be used to create *in vitro* chicken, sparing chickens—the most abused individuals used for food by virtue of their sheer number, with a million slaughtered in the U.S. alone each hour—mass suffering and death.

The suggestion that PETA's contest was a "smashing success" relies on the idea that more people were motivated to pursue this technology (through direct research or through funding) than would have been otherwise. Such impacts are nearly impossible to establish. What can be determined with more confidence is the amount of press generated by the contest for this new technology—stories appeared in major news outlets in North America (like the New York Times, ABC News, and the CBC),[36] with more online commentary, including no small amount of skepticism as to the true motives behind the prize.[37] Such press did help put the spotlight on this new mode of protein machine, at least for a while, and may have increased the likelihood that the next chicken of tomorrow will have migrated from the factory to the laboratory.[38]

It is worthwhile to note that this push from the factory into the laboratory also furthers a kind of invisibility related to protein production. That is, as we explored in A Chicken, Part 1, the move to the factory contributed to a removal of meat production from the everyday experience of many people. Operations moved to the margins of human settlement, where large numbers of birds were kept indoors. When consumers encountered chickens, it was in grocery stores or restaurants already transformed into meat—as whole carcasses, as breasts, thighs, legs, and wings, or increasingly as further processed items like nuggets and tenders. *In vitro* meat participates in this trend toward invisibility, creating an even further remove between production of meat and its animal origins. This "remove," it can be argued, is a positive development—making meat while reducing suffering. It also does increase the conceptual distance between the consumer and the food she consumes, related to production.

…WITH FLAWS?

As it stands, however, the fact of *in vitro* meat is still largely theoretical, and at this point in time, so too will be the potential flaws that may be revealed once this technology becomes real. These emergent flaws will depend upon the means by which the technology develops and the manner in which it becomes integrated into the larger socioeconomic networks, because while research groups have shown that the process is technically possible, there is a long way to go to determine whether the process will be economically viable and what precise shape it will take. As we explored earlier, many hurdles remain, but as PETA's concluding thoughts on the contest suggests, people still have a lot of hope for the process. The very possibility of the technology is reflected in its various "promissory narratives"[39]—people can tell stories about it that reflect hope for a particular type of future. That hope can be expressed by animal rights advocates and critics of industrialized meat production who promote this technological pursuit as a reasonable[40] or even necessary[41] effort to solve the problems of meat. It can be enthusiastic support for all of the positive benefits the technology might provide,[42] or a reluctant endorsement due to a lack of faith in the ability of humans to change their tastes and eating habits. If humans will not be better, then meat needs to be.[43] However, it is worth considering the extent to which this approach merely solves problems, or whether it also poses some of its own.

The future orientation of this technology makes it difficult to focus on specific flaws, because so many elements are dependent on the precise material configurations that emerge. If *in vitro* meat achieves the relevant refinement of technique and economies of scale to rival conventional protein production, what model of production and distribution would be promoted? One resembling the more contemporary food distribution related to grocery stores? One modeled on technology start-ups, with centralized control but wide distribution (along the lines of counter-top meat machines)?[44] Will *in vitro* meat be a boutique item (either because of the price or scarcity of the product, or because of a focus on rare, extinct, or unique meats)?[45] Will it instead become ubiquitous as a source of protein, perhaps even turning meat produced *in vivo* into a high demand commodity?[46] Will *in vitro* meat disrupt and thus prompt a reorganization of the existing agricultural economy? Will, ultimately, these techniques fail to develop, interest wane, and the technology fail to realize its potential? Part of the uncertainty in these potential futures stems from the need to explore a more fully realized world in which such a technology might fit. We will look more systematically at how this technology fits within a broader sense of a future in the next chapter. Here, though, given that humans are pursuing this technology and devoting resources to its promise and its potential, we would do well to reflect on some basic elements of the pursuit, and how this type of protein machine orients us to the wider world and to ourselves, as a way to anticipate specific flaws. To that end, I want to focus on those aspects alluded to in our earlier exploration of the technology: the costs associated with control and the transformation of our sense of moral accountability.

Control can be seen as a virtue, especially when considering techno-logical systems and food, so it may appear strange to include this aspect as a potential flaw. We desire safe, predictable food supplies, and seek to limit the number of uncertainties in our existing supply chains. As we explored earlier, *in vitro* meat promises much more determination related to the composition of the protein (no more unhealthy fats or inedible compo-nents) and the containment of disease (no animal to get sick, or cross-contamination to prevent while processing the whole animal body). The technique avoids some of the uncertainties of nature.[47] But as we saw with the Chicken of Tomorrow endeavor, it was precisely the efforts to exert more control on the process and the animal that led to the emergence of specific flaws. Controlling for breed and body type led to epidemiological vulnerabilities (whole flocks more susceptible to disease) and bodily break-downs (chickens unable to bear the weight of their breast muscles). Controlling for space led to waste storage breakdowns and behavioral abnormalities. Attempts to exert control on the configuration and outputs of the protein machine set the parameters for the flaws that we have come to identify as such.

We need to reckon the costs associated with efforts to control, espe-cially when *in vitro* meat technologies work to escalate our capacity to do so. Those costs are best understood in terms of the responsibilities we take on to ensure we achieve the desired outcomes. Again, the past can be instructive. Tinkering with biological processes—in order to achieve the desired chicken body, for example, or to ramp up the efficiencies of the process—requires that we have a workable understanding of how those processes function and what is necessary in order to shape them.[48] The fact of the chicken of tomorrow (and these initial forays into *in vitro* meat) is a testament to that understanding. But, to paraphrase Ruth Harrison's earlier question—do we know enough? We can identify the flaws of con-ventional protein machine as a further testament to the incompleteness of our understanding. We know enough to make it happen, but not enough to keep it from falling apart. *In vitro* meat offers an extension of that need to know. Our responsibilities shift from the need to modify bodies in order to generate the desired protein to the need to create a substitute body in order to do the same. Current struggles to replicate muscle out of basic protein represent insufficiencies that may be overcome. As we exert even more control, however, our capacity for failure may also increase.

We may also be divorcing ourselves from a sense of shared responsibility for that possible failure. *In vitro* meat technology, while reconfiguring the technique away from animal suffering, also furthers the trend toward invisibility and away from individual accountability. What our food is, and where it comes from, ends up behind a thicker veil than before. The invisibility is the product not just of physical distance, but of conceptual distance, as well, because of the specialized knowledge required to recognize the efforts and their significance. The process becomes the product of science, conducted in the rarefied space of the lab—a place even more removed from everyday experience. This sense of removal helps contribute to the perception that the problems of food (or of the environment in general) are not our problems, but the problems of experts and specialists. In the context of this technology, most people cannot contribute to the development of food in this way. Instead, we are pushed to the margins and asked just to endorse the decisions of others through consumer behavior.

This kind of technological innovation that eliminates the overt moral hazard associated with eating animals by eliminating the animal is part of the argument in favor of pursuing this lab-based meat. Giving up on the prospect of people making different choices about meat, or reforming their habits related to eating, we can instead make meat better. This ethical positioning renders humans as passive participants in the process, by offering a kind of ethical efficiency to go along with the technical one. By allowing people to act the same, while eliminating problems of production, people are compelled to act in a manner that produces better outcomes because they are not acting differently at all. The ethical questions do not get resolved as much as they fade into the design infrastructure, along with the body that contained them, leaving the role of the meat eater relatively or minimally uninterrupted. Presenting the meat eater as a kind of fixed identity, the *in vitro* meat approach allows that identity to remain intact, while changing the elements that led to the present problems. People can potentially eat the same way they do now, with fewer environmental or even personal negative health effects.

People would have to accept this product as functionally, texturally, and conceptually equivalent to meat for this to work, however. As with the chicken of yesterday, this new form of protein will have to be marketed as an acceptable substitute and people will need to get over any residual reluctance. That served as a major component of the PETA-sponsored contest, and is not a strict given. The question of adoption raises an interesting paradox within the approach—to the extent that people are fully chicken eaters, this technology is deemed necessary to satisfy those desires in a more benign manner. To the extent that people have some malleability in their diets, they are seen as willing to alter their expectations about what meat can be in order to accept the developing *in vitro* method. So, in this sense, humans are seen in this process as both fixed and fluid. What we need to recognize, and will explore more in the next chapter, is that humans are configured along with the machine. We are implicated in the process, and afforded a place in the future heralded by this potentially transformative technology. One important consideration, then, involves how we as humans are anticipated by the technology. Would we or should we accept a passive role in this system—one that assumes that we are uninterested in or incapable of handling complex moral questions?

This newer version of the protein machine touted by PETA's twenty-first century contest reflects a growing awareness of the problems associated with its twentieth century counterpart. The approach it takes, however, finds more commonality than difference. Both contests focus explicitly on the body of the chicken as a preferred site of intervention—the way to solve the problem is to make the body better. Both require a drastic revision of the supporting structures (the environments) that enable the bodies to flourish—a factory model for the Chicken of Tomorrow, a laboratory model for the "Chicken of Tomorrow." Both look to exert controls on that body and those structures in an effort to produce a specified and desired output. Both, that is, engage the problem of "meat" as a technological one, building better protein machines. They also both possess a vision of the future, ones that include not just meat but the kind of people that will consume it. The twentieth century is here—we have inherited it, and thus it is easier to identify its contours of success and failure. The *in vitro* meat future is much less certain, and will require that we imagine the kind of worlds that this type of technology might help to create. This will be the focus of our next two chapters.

NOTES

1. Unlike her previous farm life, my mother does not eat any of her current hens or roosters. My father cannot even bear the thought, which leads to a curious state of affairs. Once his chickens have come to the end of their

productive lives, he will arrange to have another individual collect them. He expected processed chickens in return, but was adamant that they not be his originally.

2. Policies vary from city to city, but many municipalities allow a strictly limited number of hens, with prohibitions against crowing birds (e.g., roosters). Of course, plenty of people skirt restrictions and raise birds anyway. My brother is compliant with the law.

3. This phenomenon is not limited to poultry. In the context of gardening, see for example, Alexander, *The $64 Tomato*.

4. Pacelle, Wayne. 2015. "Breaking News: McDonald's Announces Cage-Free Commitment for Laying Hens · A Humane Nation." A Humane Nation. September 9, 2015. https://blog.humanesociety.org/2015/09/mcdonalds-announces-cage-free-commitment-for-laying-hens.html.

5. Chipotle, "Food with Integrity." n.d. https://www.chipotle.com/food-with-integrity. Accessed May 28, 2018.

6. We will explore this approach and its limitations more in Chap. 4.

7. European Commission, "Animal Welfare on the Farm." Those freedoms include: from hunger and thirst; from discomfort; from pain, injury, and disease; to express normal behavior; from fear and distress.

8. This difference between consumer activity and collective political action will be explored more fully in Chap. 4.

9. PETA, "In Vitro Meat Production – Contest Rules." n.d. www.mediapeta.com. http://www.mediapeta.com/peta/PDF/In_Vitro_Contest_Rules.pdf. Accessed May 28, 2018.

10. See University of Pittsburgh Medical Center. n.d. "Bacon That's Good For You? Researchers Create Pigs That Produce Heart-Healthy Omega-3 Fatty Acids." ScienceDaily. Accessed May 28, 2018. https://www.sciencedaily.com/releases/2006/03/060327084435.htm. For a more recent version, see Hosie, Rachel. 2017. "Bacon May Have Just Got Healthier." The Independent. October 25, 2017. http://www.independent.co.uk/lifestyle/health-and-families/gm-pigs-less-fat-bred-scientists-genetically-modified-meat-a8018641.html.

11. Jochems, Carlo E. A., Jan B. F. van der Valk, Frans R. Stafleu, and Vera Baumans. 2002. "The Use of Fetal Bovine Serum: Ethical or Scientific Problem?" *Alternatives to Laboratory Animals: ATLA* 30 (2): 219–27.

12. Burke, Maria. 2017. "'Remarkably Flavourful' Lab-Grown Poultry." Chemistry World. March 21, 2017. https://www.chemistryworld.com/news/remarkably-flavourful-lab-grown-poultry/3007005.article.

13. The terms of the PETA-sponsored contest appear to be vague with regard to what constitutes *in vitro* meat, at least in terms of the nature of the growth medium.

14. The fish experiment and the burger proof of concept "Both agreed that the patty, prepared by chef Richard McGeown, wasn't quite what you'd get from a slaughtered cow, calling out the absence of fat as being particularly noticeable. We're still a long way, too, from being able to produce synthetic meat quickly and affordably. This morning's meal took three months to produce, which, as someone pointed out, is still 'faster than a cow.'" Abrams, Lindsay. 2013. "Here's How the World's First Synthetic Meat Tastes." Salon.com. http://www.salon.com/2013/08/05/heres_how_the_worlds_first_synthetic_meat_tastes/.

15. See, for example, Axworthy, Nicole. 2019. "Price of Lab-Grown Meat to Plummet From \$280,000 to \$10 Per Patty By 2021." VegNews.Com. July 14, 2019. https://vegnews.com/2019/7/price-of-lab-grown-meat-to-plummet-from-280000-to-10-per-patty-by-2021. For a more detailed contemporary assessment of the viability of the technology that calls into question that optimistic price point, see Fassler, Joe. "Lab-grown meat is supposed to be inevitable. The science tells a different story." *The Counter*, September 22, 2021. https://thecounter.org/lab-grown-cultivated-meat-cost-at-scale/.

16. See, for example, Wolfson, Wendy. 2002. "Lab-Grown Steaks Nearing the Menu." http://www.newscientist.com/article/dn3208-labgrown-steaks-nearing-the-menu.html.; Jones, Nicola. 2010. "Food: A Taste of Things to Come?" *Nature* 468 (7325): 752–53. https://doi.org/10.1038/468752a.; Specter, Michael. 2011. "Test-Tube Burgers." *The New Yorker*, May 23, 2011.

17. Adapted from Edelman, P.E., D.C. McFarland, V.A. Mironov, and J.G. Matheny. 2004. "In Vitro Cultured Meat Production." NewHarvest.org. 2004. http://www.new-harvest.org/img/files/Invitro.pdf.

18. It remains to be seen, in the age of *terroir* and other forms of specified or place-based food appreciation, whether this technology would seek to erase animal origins, or create quality lines associated with especially desirable animals.

19. Quoted in Pollan, *The Omnivore's Dilemma*, 97.

20. Landecker, Hannah. 2007. *Culturing Life: How Cells Became Technology.* Cambridge, MA: Harvard University Press: 232.

21. Landecker, *Culturing Life*, 72.

22. See Dennett, Daniel C. 1989. "The Origins of Selves." *Cogito* 3 (3): 163–173. https://doi.org/10.5840/cogito19893348. In this essay, he offers a thought experiment in the reading that never fails to illuminate the distinction we draw between inside and outside. First, he asks you to swallow—and thus you swallow saliva in your mouth. Now imagine spitting that saliva into a cup and drinking it.

23. Smith & Daniel, *The Chicken Book*.

24. Smith & Daniel, *The Chicken Book*. Also, consider the tale of the two pig producers who worked in different buildings on the same grounds, and had to eat dinner at separate times to avoid any chance of cross-contamination.

25. Specter, "Test-Tube Burgers," 37. In an additional emphasis on control, researchers must also be able to impose limits on how the cells divide, so they don't become what he calls "genetic miscreants."

26. The total costs associated with producing enough protein to taste came in at $325,000, amounting to around 3 patties-worth. See Fountain, Henry. 2013. "A Lab-Grown Burger Gets a Taste Test." *The New York Times*, August 5, 2013, sec. Science. https://www.nytimes.com/2013/08/06/science/a-lab-grown-burger-gets-a-taste-test.html. That same burger was described as having the mouth feel of cake.

27. As a point of reference, according to the U.S. Bureau of Labor Statistics, in December 2019, a lb. of ground beef averaged US$3.86 currently, and boneless breast meat chicken cost US$3.11 (U.S. Bureau of Labor Statistics, "Average Retail Food and Energy Prices, U.S. and Midwest Region." n.d. Accessed January 14, 2020. https://www.bls.gov/regions/mid-atlantic/data/averageretailfoodandenergyprices_usandmidwest_table.htm).

28. There is continual debate on this point, as evidenced by the controversy surrounding the Good Food Institute's techno-economic assessment. According to Fassler ("Lab-grown meat is supposed to be inevitable," 2021), the Institute's summary paints an optimistic picture, but other independent analysis of the same data reach much more pessimistic conclusions.

29. See Cannavò, Peter. 2010. "Listening to the 'Yuck Factor': Why In-Vitro Meat May Be Too Much to Digest." Washington, DC.

30. Hopkins, Patrick D., and Austin Dacey. 2008. "Vegetarian Meat: Could Technology Save Animals and Satisfy Meat Eaters?" *Journal of Agriculture and Environmental Ethics* 21: 579–96 (586).

31. Something Pollan popularized as the omnivore's dilemma.

32. Of course, part of the dilemma that we saw in Chap. 1 is precisely the ability of industrialized meat production to avoid that kind of scrutiny.

33. See, for example, Mayes, Christopher. n.d. "The Limit of Labels: Ethical Food Is More than Consumer Choice." The Conversation. Accessed January 14, 2020. http://theconversation.com/the-limit-of-labels-ethical-food-is-more-than-consumer-choice-59908.

34. See Keim, Brandon. 2008. "FDA: Don't Ask, Don't Tell on Cloned Meat." *Wired*, January 15, 2008. https://www.wired.com/2008/01/fda-dont-ask-do/.

35. Tannen, Rob. 2007. "Simplicity: The Distribution of Complexity." Boxes and Arrows. January 30, 2007. https://boxesandarrows.com/simplicity-the-distribution-of-complexity/.
36. See Schwartz, John. 2008. "PETA's Latest Tactic: $1 Million for Fake Meat." *The New York Times*, April 21, 2008, sec. U.S. https://www.nytimes.com/2008/04/21/us/21meat.html; Phillips, Ashley. 2008. "PETA Offers $1M Prize for Lab-Grown Meat." ABC News. April 23, 2008. https://abcnews.go.com/Technology/story?id=4704447&page=1; CBC News. 2008. "PETA Dishes up $1M for 'in-Vitro' Meat Contest | CBC News." CBC. April 23, 2008. http://www.cbc.ca/news/peta-dishes-up-1m-for-in-vitro-meat-contest-1.699292.
37. See Engber, Daniel. 2008. "The Bogus $1 Million Meat Prize." *Slate*, April 23, 2008. http://www.slate.com/articles/health_and_science/science/2008/04/the_bogus_1_million_meat_prize.html; Smith, Wesley J. 2008. "PETA Contest for 'In Vitro' Meat." *National Review* (blog). April 21, 2008. https://www.nationalreview.com/human-exceptionalism/peta-contest-vitro-meat-wesley-j-smith/.
38. Though it may not be based on muscle, but rather vegetable-based substances like those next burgers, etc., that more closely approximate meat.
39. O'Riordan, Fotopoulou, and Stephens. "The First Bite." See also Wurgaft, Benjamin. *Meat Planet: Artificial Flesh and the Future of Food*. Oakland, California: University of California Press, 2019.
40. See Pluhar, "Meat and Morality" and Singer, Peter. 2013. "The World's First Cruelty-Free Hamburger." *The Guardian*, August 5, 2013. http://www.theguardian.com/commentisfree/2013/aug/05/worlds-first-cruelty-free-hamburger.
41. Hopkins & Dacey, "Vegetarian Meat"; Schonwald, Josh. 2009. "Future Fillet." 2009. http://magazine.uchicago.edu/0906/features/future_fillet.shtml.
42. Saletan, William. 2006. "The Conscience of a Carnivore." 2006. http://www.slate.com/id/2142547/.
43. Deych, Rina. 2005. "How One Vegan Views In-Vitro Meat." 2005. http://www.rrrina.com/invitro_meat.htm. Accessed 22 March 2018.
44. Some source related to the model of economic disruption/tech start-ups.
45. Something like Jamón ibérico.
46. We will explore this possibility more fully in the next chapter. Speculative fiction, however, has long suggested that the emergence of cheap, plentiful synthetic goods will make more "traditional" or "natural" craft the mark of high status. See, for example, Stephenson, Neal. 2000. *The Diamond Age: Or, a Young Lady's Illustrated Primer*. Reprint edition. New York: Spectra.
47. See Hopkins & Dacey, "Vegetarian Meat."

48. For example, Kate Soper discusses the idea of freeing human reproduction from the need for gametes from opposite sexes, and notes that "we would have to know an awful lot more about biological law and process before we could even begin to commit ourselves to such a scenario." Soper, Kate. 1996. "Nature/'nature'." In *FutureNatural: Nature, Science, Culture*, edited by George Robertson, Melinda Mash, Lisa Tickner, Jon Bird, Barry Curtis, and Tim Putnam, 21–34. London: Routledge: 33.

Bibliography

Abrams, Lindsay. "Here's How the World's First Synthetic Meat Tastes – Salon. Com." Salon.com, August 5, 2013. http://www.salon.com/2013/08/05/heres_how_the_worlds_first_synthetic_meat_tastes/.

Alexander, William. *The $64 Tomato: How One Man Nearly Lost His Sanity, Spent a Fortune, and Endured an Existential Crisis in the Quest for the Perfect Garden*. Chapel Hill, NC: Algonquin Books of Chapel Hill, 2007.

Axworthy, Nicole. "Price of Lab-Grown Meat to Plummet From $280,000 to $10 Per Patty By 2021." VegNews.com, July 14, 2019. https://vegnews.com/2019/7/price-of-lab-grown-meat-to-plummet-from-280000-to-10-per-patty-by-2021.

Burke, Maria. "'Remarkably Flavourful' Lab-Grown Poultry." Chemistry World, March 21, 2017. https://www.chemistryworld.com/news/remarkably-flavourful-lab-grown-poultry/3007005.article.

Cannavò, Peter. "Listening to the 'Yuck Factor': Why In-Vitro Meat May Be Too Much to Digest." Washington, D.C., 2010.

CBC News. "PETA Dishes up $1M for 'in-Vitro' Meat Contest | CBC News." CBC, April 23, 2008. http://www.cbc.ca/news/peta-dishes-up-1m-for-in-vitro-meat-contest-1.699292.

Chipotle. "Food with Integrity," May 28, 2018. https://www.chipotle.com/food-with-integrity.

Dennett, Daniel C. "The Origins of Selves." *Cogito* 3, no. 3 (1989): 163–73. https://doi.org/10.5840/cogito19893348.

Deych, Rina. "How One Vegan Views In-Vitro Meat," 2005. http://www.rrrina.com/invitro_meat.htm. Accessed 22 March 2018.

Edelman, P.E., D.C. McFarland, V.A. Mironov, and J.G. Matheny. "In Vitro Cultured Meat Production." NewHarvest.org, 2004. http://www.newharvest.org/img/files/Invitro.pdf.

Engber, Daniel. "The Bogus $1 Million Meat Prize." *Slate*, April 23, 2008. http://www.slate.com/articles/health_and_science/science/2008/04/the_bogus_1_million_meat_prize.html.

Edgerton, David. *The Shock of the Old: Technology and Global History since 1900*. Oxford: Oxford University Press, 2007.

European Commission. "Animal Welfare on the Farm – Food Safety – European Commission." Food Safety, May 23, 2018. https://ec.europa.eu/food/animals/welfare/practice/farm_en.

Fassler, Joe. "Lab-grown meat is supposed to be inevitable. The science tells a different story." *The Counter*, September 22, 2021. https://thecounter.org/lab-grown-cultivated-meat-cost-at-scale/.

Fountain, Henry. "A Lab-Grown Burger Gets a Taste Test." *The New York Times*, August 5, 2013, sec. Science. https://www.nytimes.com/2013/08/06/science/a-lab-grown-burger-gets-a-taste-test.html.

Hopkins, Patrick D., and Austin Dacey. "Vegetarian Meat: Could Technology Save Animals and Satisfy Meat Eaters?" *Journal of Agriculture and Environmental Ethics* 21 (2008): 579–96.

Hosie, Rachel. "Bacon May Have Just Got Healthier." The Independent, October 25, 2017. http://www.independent.co.uk/life-style/health-and-families/gm-pigs-less-fat-bred-scientists-genetically-modified-meat-a8018641.html.

Jochems, Carlo E. A., Jan B. F. van der Valk, Frans R. Stafleu, and Vera Baumans. "The Use of Fetal Bovine Serum: Ethical or Scientific Problem?" *Alternatives to Laboratory Animals: ATLA* 30, no. 2 (April 2002): 219–27.

Jones, Nicola. "Food: A Taste of Things to Come?" *Nature* 468, no. 7325 (December 9, 2010): 752–53. https://doi.org/10.1038/468752a.

Keim, Brandon. "FDA: Don't Ask, Don't Tell on Cloned Meat." *Wired*, January 15, 2008. https://www.wired.com/2008/01/fda-dont-ask-do/.

Landecker, Hannah. *Culturing Life: How Cells Became Technology.* Cambridge, MA: Harvard University Press, 2007.

Mayes, Christopher. "The Limit of Labels: Ethical Food Is More than Consumer Choice." The Conversation. Accessed January 14, 2020. http://theconversation.com/the-limit-of-labels-ethical-food-is-more-than-consumer-choice-59908.

O'Riordan, Kate, Aristea Fotopoulou, and Neil Stephens. "The First Bite: Imaginaries, Promotional Public and the Laboratory Grown Burger." *Public Understanding of Science* 26, no. 2 (2017): 148–63.

Pacelle, Wayne. "Breaking News: McDonald's Announces Cage-Free Commitment for Laying Hens A Humane Nation." A Humane Nation, September 9, 2015. https://blog.humanesociety.org/2015/09/mcdonalds-announces-cage-free-commitment-for-laying-hens.html.

PETA. "In Vitro Meat Production – Contest Rules." www.mediapeta.com, May 28, 2018. http://www.mediapeta.com/peta/PDF/In_Vitro_Contest_Rules.pdf.

Phillips, Ashley. "PETA Offers $1M Prize for Lab-Grown Meat." ABC News, April 23, 2008. https://abcnews.go.com/Technology/story?id=4704447&page=1.

Pluhar, Evelyn B. "Meat and Morality: Alternatives to Factory Farming." *Journal of Agriculture and Environmental Ethics* 23, no. 5 (2010): 455–68.

Pollan, Michael. *The Omnivore's Dilemma: A Natural History of Four Meals.* New York: Penguin Press, 2006.

Saletan, William. "The Conscience of a Carnivore," 2006. http://www.slate.com/id/2142547/.

Schwartz, John. "PETA's Latest Tactic: $1 Million for Fake Meat." *The New York Times*, April 21, 2008, sec. U.S. https://www.nytimes.com/2008/04/21/us/21meat.html.

Smith, Page, and Charles Daniel. *The Chicken Book.* Boston: Little, Brown, and Company, 1975.

Smith, Wesley J. "PETA Contest for 'In Vitro' Meat." *National Review* (blog), April 21, 2008. https://www.nationalreview.com/human-exceptionalism/peta-contest-vitro-meat-wesley-j-smith/.

Soper, Kate. "Nature/'nature'." In *FutureNatural: Nature, Science, Culture*, edited by George Robertson, Melinda Mash, Lisa Tickner, Jon Bird, Barry Curtis, and Tim Putnam, 21–34. London: Routledge, 1996.

Specter, Michael. "Test-Tube Burgers." *The New Yorker*, May 23, 2011.

Stephenson, Neal. *The Diamond Age: Or, a Young Lady's Illustrated Primer.* Reprint edition. New York: Spectra, 2000.

Tannen, Rob. "Simplicity: The Distribution of Complexity." Boxes and Arrows, January 30, 2007. https://boxesandarrows.com/simplicity-the-distribution-of-complexity/.

University of Pittsburgh Medical Center. "Bacon That's Good For You? Researchers Create Pigs That Produce Heart-Healthy Omega-3 Fatty Acids." ScienceDaily, May 28, 2018. https://www.sciencedaily.com/releases/2006/03/060327084435.htm.

U.S. Bureau of Labor Statistics. "Average Retail Food and Energy Prices, U.S. and Midwest Region: Mid–Atlantic Information Office." Accessed January 14, 2020. https://www.bls.gov/regions/mid-atlantic/data/averageretailfoodandenergyprices_usandmidwest_table.htm.

Wolfson, Wendy. "Lab-Grown Steaks Nearing the Menu," 2002. http://www.newscientist.com/article/dn3208-labgrown-steaks-nearing-the-menu.html.

Wurgaft, Benjamin. *Meat Planet: Artificial Flesh and the Future of Food.* First edition. Oakland, California: University of California Press, 2019.

CHAPTER 5

A Future, Part I

Abstract This chapter explores the fascination with a chicken-less chicken future, as portended by these current *in vitro* efforts, through the lens of speculative fact and fiction. The narrative starts with an era of expansionist dreams of limitless chicken flesh found in the early and mid-nineteenth century. It then switches to a stronger focus on design, where we can exercise our intentions more precisely on protein machines and their outputs. These fictional accounts and factual promotions each explore different ways in which the problems of meat have been "solved," and echo much of the discussion surrounding the development and promotion of *in vitro* meat technology, while informing a society that would support it.

Keywords Science fiction • Speculative future • Expansion • Design

> *Works of art cannot save us. They can simply render us more sensitive to what needs to be repaired.*
> —Terry Eagleton, *Reason, Faith, and Revolution: Reflections on the God Debate*, New Haven, CT: Yale University Press, 2010

W. Galusky, *Protein Machines, Technology, and the Nature of the Future*, https://doi.org/10.1007/978-3-031-08717-2_5

As we saw in the last chapter, *in vitro* meat technologies set their sights on a future that has not yet come to pass, and this fact complicates any attempt to reckon fully with the world such a technology might help to create. In analyzing the mid-twentieth century Chicken of Tomorrow contest, we have the fact of our present and what resulted when the vision became real. This reality enables the recognition of the accomplishments and the flaws that emerged with this configuration of a protein machine. These features are emergent—not fully anticipated in design and in implementation, but also the result of the unforeseen and the overlooked. This state of things leads to our present dilemma. On the one hand, we can confront those problems that have been made real in our current world. On the other, visions of new interventions and solutions will have their own emergent properties that by design cannot be fully anticipated. The future, as they say, is unwritten. It is uncertain precisely how things will unfold or how much control we will be able to exert over the processes and consequences of implementation. We do not know if the problems we encounter will be worth the effort it takes to change our current practices.

When I first got chickens, I had a simple dream involving fresh eggs made possible by cute animals running around in the yard. This was a means of correcting a problem—my own disconnection with food and the contemporary means of producing it. My encounter with chickens was not quite so simple, of course, and the direction my life took was altered quite meaningfully. Plans to use the existing hen house at my rented home? Thwarted. Plans to collect eggs from the nesting boxes? Thwarted. Plans to have *hostas* in front of the house that did not look like Swiss cheese, or shoes that did not have beak marks?[1] Thwarted. Plans to keep eight chickens for a good long time? Thwarted.[2] The hens quickly abandoned the henhouse, laid their eggs in obscure and hard to locate areas, shredded the *hostas* and scuffed a fair number of shoes, and many chickens died or disappeared. Once the beleaguered landscape of those chaotic initial months settled, I had a lone chicken in a small coop spending her days with two goats—not the agricultural idyll I had envisioned, but one that I could eventually come to accept. I was willing to adapt to this relatively unplanned reality, in large part because so little was at stake (a hobby, not a livelihood; a lark, not a lifeline). I felt more prepared the second time around. The truth is, however, even following what I learned from that initial experience and starting to set out on a new one with what I felt was a more realistic expectation of what a future with chickens would look like, my new flock still managed to redirect me. This flock of varying number did not respect territorial boundaries (necessitating a fence, and then a better fence) and even managed to lay an egg on the floorboard of my car (necessitating rolled up windows, even in the garage). What seems true is that my sense of the future is in constant negotiation with theirs, and we end up in a new place together.

My experiences here are consistent with arguments made by Bruno Latour and Peter-Paul Verbeek, among others,[3] related to the interplay between technologies and users—as objects get used, they get altered, but so do the people who use them. We will dive into this relationship more thoroughly in the next chapter, but for now I want to reflect on how this insight complicates even further any prognostication of the future, especially with the introduction of a new technology designed to fix a present problem. At the very least, we should not simply assume that a technology like *in vitro* meat will change nothing except existing meat production technologies—that everything else stays the same, including socioeconomic relationships, attitudes, and behaviors, except for the way meat is made. To imagine a change in technology into the future requires that we imagine a changed world, as well. One that might accept or reject or, better yet, reflect the focus of that new technology. It would be shortsighted of us to imagine that as the technology changed and *in vitro* meat became possible—to the extent that PETA's contest required, in fact—the rest of humanity and its structures stayed exactly the same. Just as the Chicken of Tomorrow contest inaugurated a change in how chicken is made, it also participated in a much wider transformation, in agricultural practice, land use, transportation, human settlement patterns, and a shifting understanding of diet and nutrition. Tomorrow was not just about a new chicken, but a new chicken eater in a new world.

It is also important to acknowledge that *in vitro* meat may never come into wide scale existence as a true alternative to more conventional (animal-based) meat, or transform agricultural production to any appreciable extent. Plenty of factors might derail current efforts—economic, political, technical—that would enable the difficult transition from possible to actual. Warren Belasco, in his book *Meals to Come*, explores a century's worth of future food predictions, most of which are confined to the dustbins of history and provoke a kind of knowing smirk now. He notes that these predictions are products of their time, often reflecting self-interested individuals and groups looking to shape (and often profit from) the future. Nevertheless, he argues, studying past visions of the future of food with the right mindset (what he terms a "healthy skepticism") enables us to be grounded in the realities that need to be faced, but also to "[allow] our-selves to dream of a better future."[4] It may be a future that merely averts catastrophe, or offers a utopian vision of an improved world. To that end, *in vitro* meat discourse is involved in this type of formulation—offering a particular vision of an improved world currently served by confinement agriculture.

In this chapter, I want to do two related things. I want to explore a his-tory of the discourse surrounding the dream of animal-less meat, and I want to do so through narratives that situate this technology (as much as possible) in a fuller, transformed world. Real historical events lead to fic-tional explorations of their meaning, their potential, and their threat. Starting in the early twentieth century, the possibility of animal-less meat has been a part of scientific and popular discourse. On the "real" side, I start with explorations of the work of Alexis Carrel, who claimed to have kept a single piece of fetal chicken heart tissue alive for decades. Much later in that century, breakthroughs in cell culturing techniques and genetic engineering led to real optimism that tissue cultures could be cul-tivated and controlled, and thus more fully reflect human intention and design. Both moments fed the imagination of dramatists, technological optimists, and novelists. Drawing on those early efforts to keep chicken cells alive in culture, I look at works of fiction (a radio drama and a novel) that tried to take the idea and, quite literally, expand on it. As the fervor of the cultured chicken heart gave way to actual achievements in biological plasticity, I then take up Margaret Atwood's speculative fiction trilogy starting with *Oryx & Crake*, exploring a world by design. These science fictional accounts of animal-less meat each explore different ways in which the problems of meat have been solved and, importantly, link the develop-ment and promotion of this new technology to the priorities of a society that would support it. To what extent can we imagine a technology and a society that change together? How might these stories help us more fully realize how technologies come to reflect the worlds that contain them?

A Century of Science Fact and Fiction

One way to explore our possible futures involves looking to the past, just as we did with the mid-century chicken of tomorrow contest. That explo-ration helped to reveal that the reality of the protein machine contains more than its initial vision, suggesting that explorations of our tomorrow would do well to consider more than just the technical aspects of our new configuration. We can also look to the past to find attempts to create lab-based protein and to reflect on the ways in which these efforts prompted new narratives of possibility. The idea of chicken (protein)—chicken (body) = better chicken has been around for over a century, spurred on by concrete innovations, which are quickly followed both by dreams of a technophilic future and a growing sense of disquiet. That, in fact, tends to be the order of things—some publicized scientific success story sparks the imagination of the wider public who begin to explore the utopic and dystopic possibilities of this innovation. For our focus, on protein machines that do not rely on animal bodies, we can find two specific flurries of activity, starting all the way back to the early years of the twentieth century. The breakthroughs of those moments led to different forms of reflection, leading to what I have taken to calling an age of expansion and an age of design. The former fixates on unbridled growth once we viewed cell death as no longer inevitable, as one of a few "preventable occurrences." The latter reflects a greater focus on control, not just unleashing life but shaping it to our particular desire. Both focus on a key and developing relationship to biological entities, and use that relationship as a basis for generating hope for, and fear of, the future.

An Age of Expansion

In the age of expansion, we should first consider the real.* I will need to explain that asterisk below. Alexis Carrel, in the early stages of the twentieth century, asked the question—freed from its animal body, its natural container, *in vivo*, are there any other natural checks on the growth of the cell? He argued, and deigned to show experimentally, that any such checks could be overcome with the appropriate technique. Freed of the body or, as Carrel puts it, "liberated from the control of the organism from which they were derived,"[5] cells might exhibit permanent life and grow indefinitely. His experimental protocol involved using fetal heart tissue from a chick, grown in solution. According to the results published by Carrel in 1912 and later, and carried out by associates into the middle part of that same century, this original starter cell culture was touted to grow in perpetuity, excepting any artificial constraints imposed by research protocols and lab specifications.

For Carrel, the body operates as a control on cell growth—after all, uncontrolled, runaway cell growth within a body is widely acknowledged to be a pathology. Bodies function to regulate cell growth, unless something goes wrong. Perhaps, though, removed from this natural context, cells could exhibit fewer limitations. In his paper, "On the Permanent Life of Tissues Outside of the Organism," Carrel suggests that cell cultures removed from bodies may be able to escape their typically short life. Rather:

> It might be supposed that senility and death of the cultures, instead of being necessary, resulted merely from preventable occurrences; such as accumulation of catabolic substances and exhaustion of the medium. The suppression of these causes should bring about the regeneration of old cultures and prevent their death. It is even conceivable that the length of the life of a tissue outside of the organism could exceed greatly its normal duration in the body, because elemental death might be postponed indefinitely by a proper artificial nutrition.[6]

Such occurrences, then, the ones tied most closely to aging and death, are potentially preventable.[7] One might see the obvious and profound implications for humans and their own biological precariousness. Carrel's work with fetal chicken heart tissue was symbolically important as well—one piece of evidence that was able to demonstrate the continued living status of these cells was its capacity to "pulsate rhythmically."[8] This work might pave the way for protocols that could harness the natural and potentially boundless potential of cells and stop aging and even death. And, as people hypothesized later, maybe even feed people.

It should come as no surprise that this research caught the fancy of the press over the decades. The focus on this enlarging mass of chicken flesh, begun from a single sample, led to innumerable articles, profiling Carrel and offering metaphorical images meant to capture the potential enormity of the chicken heart. Publications seemed to strive to outdo each other. One newspaper of the time described the phenomenon to its readers this way: "If all of the cells produced by Dr. Carrel from his cultured chicken heart were kept together, they would produce a rooster that could cross the Atlantic Ocean in a single stride."[9] Other descriptions involved imagining a weathercock perched on the earth, and suggested that the potential mass of the ever expanding mound of flesh, if not managed in the lab, could potentially rival that of the sun.[10] To that end, the body boundary once removed offered the potential of cells to grow unabated and be a kind of limitless substance—perhaps matched only by human appetites.

The reality of Carrel's experimental work was eventually called into question—hence the asterisk above. Most notably, the researcher Leonard Hayflick noticed in his own work in the 1960s that he could not get his cell cultures to grow indefinitely. Instead of simply assuming his own protocols were flawed, Hayflick instead theorized and experimentally demonstrated that there is a limit to how many times a cell will divide in culture. This claim directly contradicted the notion of immortality promoted by Carrel, and led to questions about how Carrel's original cell line had been maintained. Theories include accidental contamination by embryonic fluid used as the medium (which renewed the cell culture periodically) or intentional additions of new cell tissue in order to keep the "immortal line" alive.[11] Whatever the reasons, Carrel's claims about the immortality of his cell line have since been rejected by the scientific community.

As with many profound scientific findings, however, the initial enthusiasm over imagined new possibilities far outlasted the later fall from grace. This kind of overenthusiasm is also true and somewhat typical for the scientific community, which left many to question why it took so long for the problems with Carrel's research to be exposed.[12] Nevertheless, Alexis Carrel's work captured the popular imagination, and infiltrated human conceptions both through anticipation about the possible (solving problems) and unease about the unknown (creating problems). The very idea—of chicken, grown in perpetuity, in a lab, with unchecked growth—fed visions of utopia and dystopia alike.

One person who foresaw the utopian possibility related to this expansive cell culture was Winston Churchill. "Foresaw" being the operative word here, because Churchill merely envisioned a world that would contain this kind of technology, applied to the production of food. In an essay he wrote for *Strand Magazine* in 1931, entitled "50 Years Hence," Churchill laid out his vision for the future on a number of topics, including food. He, like Carrel, also saw the animal body as something to circumvent, though less in terms of limits and more in favor of efficiency. Churchill wrote, "We shall escape the absurdity of growing a whole chicken in order to eat the breast or wing, by growing these parts separately under a suitable medium."[13] The links of influence between his work here and that of Carrel's work seem strong, in part because of the focus on chicken in this prognostication, which was not a widely consumed protein compared to others. And there is also the absurdity of the animal body—something that ideally should not be grown to just eat the meat. Edible substances are important, and inedible ones are mere inefficiencies.

Key for the present analysis is the idea that a vision of the food of the future is congruent with a broader view of that future. That is, it is not just the technologies that will be different—we will be different, too. Churchill's vision includes a broader disappearance of agriculture from human purview, for example. He forecasts a world where agricultural production will literally go underground: "food will be produced without recourse to sunlight. Vast cellars in which artificial radiation is generated may replace the cornfields or potato-patches of the world. Parks and gardens will cover our pastures and ploughed fields." The world will be more fully resemble humanity's vision for it, exercised by human desire and mechanisms of control. The means to produce such an idyll will be left invisible to the common person.[14] Churchill does not paint a purely rosy picture—he assumes a similar level of manipulation will be applied to humans ourselves, and the vast accumulation of power that this portends can lead to ability that outstrips wisdom or morality.[15] Overall, however, Churchill views a future where the world, including how we eat, more fully resembles specific human desire, and represents the articulation and manipulation of enormous power over the world and over ourselves, which will largely be hidden from view.

A more dystopic, though simplistic, imagining of a world that contains an ever expanding, bodiless chicken can be found in a 1937 episode of the Arch Oboler radio show, *Lights Out*, entitled, "Chicken Heart." This drama's inspiration comes directly and unequivocally from Carrel's work, as evidenced by Oboler's introduction to the piece: "Do you remember some time ago, in an Eastern scientific institution? They kept a piece of heart alive for weeks on end? Well, I got to thinking. What if that heart began to grow, and grow, and grow, and grow...."[16] The story itself unfolds over the course of about seven and a half minutes, and starts as an unknown mass of flesh is bursting out of a building. No longer contained by a body, not able to be contained by a building, this pulsating mass seeks to fill whatever space it's given, and (just ever so slightly deviating from reality) to feed on whatever substance it touches. As one character helpfully describes to the listening audience at home, "This tissue is doubling in size every hour. Free in the street. And then those... those tentacles of protoplasm stretching out to feed on anything they can reach." Panic ensues, while the sense of dread for the audience is heightened by the pulsating heart rhythm that pervades the soundtrack.

There is one character who knows of the flesh's origins and its devastating potential, Dr. Alberts. He informs the government exactly what it is that seems to be threatening humanity: "that great ever-growing mass of flesh. It is, or it was, a chicken heart. [exclamations of disbelief] I tell you, that mass of flesh was a chicken heart. The tissue of which, for some reason, is undergoing constant rapid accelerating growth." But Dr. Alberts is condemned to the fate of Cassandra, as his warnings are not heeded in time, and this "gray blanket of evil" soon covers the whole earth, choking out life save for two people circling it in a plane—Dr. Alberts and a companion, Mr. Lewis. Right before that plane runs out of gas, plunging into the flesh-covered Earth with a wet splat, Dr. Alberts offers an elegy for humanity:

> DR. ALBERTS: The end has come for humanity. Not in the red of atomic fusion. Not in the glory of interstellar combustion. Not in the peace of white, cold silence. But with that, that creeping grasping flesh below us. It is a joke, eh, Lewis, a great joke. The joke of the cosmos. The end of mankind.
> MR. LEWIS: Why?
> DR. ALBERTS: Because of a chicken's heart.

In this stylized rendering, there is no future that contains bodiless meat, because the meat makes the Earth its body. Humans unleashed this harbinger of doom by unshackling it from its conventional boundary. Unleashing that flesh from its natural controls, humans found that the could not replace them, leading to the catastrophic end.

A bit more nuanced engagement with this expansionist dystopia can be found in the 1953 novel, *The Space Merchants*, by Frederik Pohl and C.M. Kornbluth. The world depicted in this book is one where consumption and consumerism determine the order of things and define the purpose of existence. People are categorized in terms of system function, with star-class copywriters (essentially top-tier advertising people) being at the pinnacle. Thus, the reader meets the protagonist, Mitchell Courtenay, star-class copywriter extraordinaire, struggling to find new markets to exploit in a world completely colonized by consumption, and with a worldwide ecosystem overtaxed and overburdened. This system is structured by the copywriters, supported by the consuming masses, and challenged by the "Consies," a fringe group who have the audacity to suggest that the world could be otherwise. The Consies are less oriented around consumerism and more so around conservation. For those in power, the Consies are more of a nuisance than a threat, and so the copywriters busy themselves with a new space, a new untapped potential market—a human colony on Venus.

Internal corporate machinations and foul play lead to the fall from grace (and station) for Courtenay, and he finds himself with a new name (George Groby) and a new social class. In a world defined by how you consume and how you produce, Courtenay finds himself not the purveyor of advertisement, but its target. And the escape he promoted in his work as a starry-eyed copywriter becomes a material necessity given the realities of his new life—not one that tells a person what to eat, but actually helps produce that food. More to the point, Courtenay finds himself face to "face" with Chicken Little.

Chicken Little is a large, ever growing mound of chicken flesh, sliced regularly and fed to the masses. As an entity, Chicken Little appears to be a direct, imagined descendent of Carrel's work, as well, being identified as an ever-expanding mass of flesh that started as heart tissue. It was a difficult creature to love, though Herrera, another character identified as a master slicer who ends up taking the protagonist under his wing, expresses a measure of appreciation. He has to almost guiltily admit that he likes the godforsaken creature. But most importantly, Herrera played an indispensable part in keeping this expansionist creature from overwhelming its confines:

> He had more than a production job. He was a safety valve. Chicken Little grew and grew, as she had been growing for decades. Since she had started as a lump of heart tissue, she didn't know any better than to grow up against a foreign body and surround it. She didn't know any better than to grow and fill her concrete vault and keep growing, compressing her cells and rupturing them. As long as she got nutrient, she grew. Herrera saw to it that she grew round and plump, that no tissue got old and tough before it was sliced, that one side was not neglected for the other.[17]

This vision of an expanding mound of chicken protein—a kind of perpetual motion protein machine—was kept in check by the attentions and activities of a group of people charged with the task of cutting steaks off of the mound for consumption. Chicken Little was a food source—a way of feeding the ever-expanding population of the world, with an ever-expanding mass of protein. Herrera and other slicers produced a quota of "steaks" that were to be shipped out across the planet, to be eaten for breakfast and dinner. It is also important to note that this product is not a highly desired one, despite its futuristic design and advanced technique. Rather, those steaks are the primary staple of those in the lower classes who cannot afford "real" food—base consumers, of the type Courtenay has become and with whom he now associates in his fallen state.

As an organism, Chicken Little had to be managed—not just in terms of its overall growth, but the type of that growth. For example, Herrera, using his preferred sobriquet for the being, *gallina,* describes the vigilance that a slicer has to have: "'Sometimes a *Gallina* goes cancer.' He spat. 'No good to eat for people. You got to burn it all if you don't catch it real fast and—' He swung his glittering slicer in a screaming arc to show me what he meant by 'catch.'" The being has to be monitored to ensure that it produces that kind of protein expected, and kept clean from the unexpected. But Chicken Little is not just seen as a necessary evil, but as a provider—of food, surely, but oddly enough also cover for subversive activities. Chicken Little as an organism is invisible to most people (like Mitchell Courtenay prior to his fall), and also helps to make certain clandestine activities invisible, as well. Herrera, for example, using a whistle, can get Chicken Little to create a pocket that allows for people to plot against the prevailing system undetected. The Consies have an office hidden beneath the depths of this pulsating protein blob.

The key for our analysis here is not the plot of the novel (which itself involves a plan to commandeer the Venus colony ship and turn it into a ticket to a Consie paradise), but rather the relationship between the technology (Chicken Little) and the society that contains it. Chicken Little is depicted as unrelenting in its desire to grow, even to its own detriment—it might develop a cancerous spot that ends up taking over the entire organism and ending in its own destruction. The creature is bound by an artificial body, a new structure that strains to hold it, and that creature is managed by a few people (and invisible to most others). Many of the same characteristics apply to the human society—driven by an unrelenting desire to consume, event to its own detriment. Straining at the boundaries of the structure (in this case, the entire Earth) that contains it. And managed by a small class of people who attempt to pull levers of desire and direct people's impulses toward predetermined ends.[18] The "space" of *The Space Merchants* does not just refer to outer space, but of space on the earth. Specifically, there is no space—there is overcrowding and references to Malthus' concerns about population growth. No more space for people, no more space for markets. Chicken Little exists in this context as a symbol of this overconsumption, this saturation—something that can be grown when nothing else can. And the entire planet is bursting at the seams.

It is important to also note that this society—one that would contain Chicken Little—is one that embraces the technological fix to complex environmental problems. This attitude is initially embodied by Courtenay prior to his fall from grace. Reflecting on the "absurdity" of the notion that nature might impose limits on human action, he notes:

> Science is *always* a step ahead of the failure of natural resources. After all, when real meat got scarce, we had soya burgers ready. When oil ran low, technology developed the pedicab.
>
> I had been exposed to Consie sentiment in my time, and the arguments had all come down to one thing: Nature's way of living with the *right* way of living. Silly. If "Nature" had intended us to eat fresh vegetables, it wouldn't have given us niacin or ascorbic acid.

A society that believes in technological fixes, which does not desire (or is unable) to correct underlying unsustainable behavior, is the kind of society that would accept and even welcome Chicken Little. Both appear focused on consumption and growth as ends in themselves, even if that means that their respective containers are outgrown.

The age of expansion in this era of science fact and fiction focused on a singular construct—the growth of the organism, freed from the body and associated "occurrences," like cell senescence and death. This kind of organism seems compatible, in Pohl & Kornbluth's telling, with a society with a similarly insatiable appetite, for consumption in its own sake, regardless of quality or utility. In both forms of expansion, this drive for excess would be detrimental in the long run. And while control was typically asserted by a few, that control was less actual and more illusory.

An Age of Design

After this age of expansion comes what might more aptly be called an age of design (although an alternative title might be an age of plasticity). The change occurs in terms of a focus less on unleashing biological potential (freed from the constraints of the body) and more in tinkering with biological possibility. That is, the biological may come to be understood not as destiny but rather as the opening bid in a negotiation over the final product. How might we understand a world that could, at the biological level, be more fully designed to human specifications? How might humans see themselves in this new milieu?

The science behind this new age is something that we explored in part in the previous chapter, as it enables the engineering and technique behind current *in vitro* meat efforts. Researchers were able to develop techniques that allowed cell cultures to be suspended in time, or made to divide simultaneously, for example, based on human intentions or desires.[19] These techniques put the mechanics of cell cultivation under human influence, altering our understanding of what biology requires and what it can reflect. We can now, for example, create fertilized human embryos in a lab, bypassing more customary modes of fertilization—essentially manufacturing an artificial womb for human conception. As Landecker notes, this kind of power reflects a more profound change to our understanding of the biological world—it does not just change what it means to be human, but rather it "changes what it means to be biological".[20] Landecker explores the breakthroughs and innovations, like cell synchrony and cryobiology, that have enabled such changes in the plasticity of biological processes. These abilities are joined by genetic modification—by mapping various genomes and harnessing gene-splicing capacities.

Humans have, of course, always influenced biological processes—for our current narrative, the notion of selective breeding[21] comes into play. The Chicken of Tomorrow was the product of this technique, for example, and domestication generally occurs though the intersection of biological mechanisms and human practices. But this development at the level of cell culture, while furthering the intention of directing biological processes, changes and refines the mechanism. Not limited to bodies (again) but tinkering with their constituent parts, and not simply removing the body as a governor to growth (as Carrel claimed to have done) but changing what might become a body. And chickens have been at the forefront of this research with agricultural food animals, because they are comparatively shorter-lived, cheaper, and quicker to reproduce than their compatriots (though, once "customary" bodies are eliminated from the equation, these constraints lose their impact).

Genetic tinkering allows for more purposeful design—creating animals with directed traits similar to selective breeding, but also generating combinations previously not possible. These capacities for design are extended through the ability to decode and repurpose genetic sequences, allowing for the elimination of some traits and the splicing of others into novel places. As such, humans have the capacity to treat genetic pathologies or create chimeras like rabbits that glow in the dark due to spliced biolumi-nescence traits and strawberries that are frost-tolerant thanks to genes har-vested from fish. Or, in another context, animal-less meat that contains precisely what we want, and nothing that we do not want. We have switched from unleashing nature to directing and designing it.

As we explored in A Chicken, Part 2, there has been no shortage of boosterism. The potential of *in vitro* meat has attracted a lot of support from a variety of entities and individuals concerned about the problems associated with meat production and human over-consumption. Groups that have sought to pursue and promote the technology include New Harvest ("Building the Field of Cellular Agriculture"), which has pursued a broad range of initiatives, including "milk without cows, eggs without hens, [and] beef without cows."[22] On this latter point, the group is a sponsor of Mark Post's company, Mosa Meats ("The Meat Revolution"),[23] pursing the *in vitro* meat burger. A company in the U.S. is Upside Foods (previously Memphis Meats—"Better meat, better world"), whose slogan is "To satisfy our cravings, our conscious, and our heart. It's easy to be on the same side when we look for the Upside."[24] Larger conventional pro-tein producers do not want to be left behind. Tyson Foods, Inc., for exam-ple, launched a venture capital fund in 2016, with the goal of making "investments in promising entrepreneurial food businesses that are pio-neering breakthrough products and technologies, as well as disruptive business models."[25] Their portfolio includes an investment in the plant-based meat alternative company, Beyond Meat ("The Future of Protein"),[26] as well as the cultured meat companies Future Meat Technologies ("Animal Free, Earth Friendly")[27] and Upside Foods.

Additionally, there has been a lot of sensational coverage on the next new thing in the future of food. Here is just a sampling of the news outlets and article titles exploring the possibility of technology, moving from questions to declarations:

- CNN (2009): "In-vitro meat: Would lab-burgers be better for us and the planet?"[28]
- The Atlantic (2013): "Is Lab-Grown Meat Good for Us?"[29]
- The Washington Post (2016): "Lab-grown meat is in your future, and it may be healthier than the real stuff."[30]
- NBC News (2017): "Lab-Grown Meat May Save a Lot More than Farm Animals' Lives."[31]

- Futurism.com (2017): "Lab-Grown Meat Is Healthier. It's Cheaper. It's the Future."[32]

These articles tend to focus on the purity of the food—what Memphis Meats labels "clean." This purity is biological (as the meat escapes exposure to pathogens and disease), as well as ethical (as the production methods appear to bypass the moral complexities associated with killing animals). The articles also emphasize eliminating the inefficiencies and emissions linked to conventional production methods. We can eat meat that has been cleansed of impurities (disease and ethics), inefficiencies (growing only what we want), and environmental degradations (associated with wastes, gas emissions, etc.).

The future of meat consumption, then, is one where human tastes for meat can be satisfied by/through substitution—a protein produced through better, cleaner processes. Humans are portrayed as willing to partake in this food produce, especially and to the extent that it is sufficiently similar to its conventionally produced partner. The tendency of this technological boosterism regarding *in vitro* meat is anchored by this sense of substitution—that it offers a better way to continue to live as we do now. We can consume meat at current levels, and even increase them, but now with fewer costs associated with such consumption: less suffering for animals, fewer health concerns for humans, less ecological stress compared with conventional modes of production. Overall, these positive views of a future of *in vitro* methods identify the means by which production changes, and humans largely get to stay the same (at least from the consumer end).[33] This relative unobtrusiveness is a large part of its presumed value, its promise, which led to PETA's contest in the first place—tapping into and promoting that promise. Though we would also need to explore the likelihood that such promise could be fulfilled—to what extent is such benign substitution possible? How might we come to understand the changes in the wider society that such a technology might itself reflect or promote?

Others are less sanguine about the world that technologies like *in vitro* meat portend in the future. People have called into question the idea that *in vitro* meat would actually serve as a benign substitute—both in terms of impact[34] and in terms of consumption.[35] In fact, as Mattick et al. note, because of the uncertainties associated with the technique and the different trajectories it might take, most likely "*all* projections of specific environmental impacts are almost certainly wrong, and it may be years before we know how far off we are."[36] The best we may be able to do, in keeping with the present analysis, is explore some of the philosophical commitments of the technology that intersect with broader potential sociotechnical changes. For that, I want to focus on one set of books which explore a world designed by humans (though not, inevitably, for humans)—the speculative fiction trilogy by Margaret Atwood, starting with *Oryx & Crake*. These books follow a world ravaged by the designs of a few presuming to create a better world for the many. This "better" world falls apart in short order, because of how shortsighted those improvements proved to be, and how quickly events slipped beyond anyone's control. Importantly, Atwood is telling a story about a possible future, but grounds it in the present. She ends *MaddAddam*, the last book in the trilogy, with the following caveat: "Although *MaddAddam* is a work of fiction, it does not include any technologies or biobeings that do not already exist, are not under construction, or are not possible in theory."[37] The prospects that Atwood explore are not inevitable, but are possible, and reflect such a possibility because of how we interact with biology—as something that is plastic and may more fully reflect human intentions.

Atwood's trilogy traces the dissolution of the existing world, and its subsequent reorganization and recalibration along differing lines. The fundamental catalyst for the cataclysm is a single individual, who calls himself Crake. A design bio engineer for a large conglomerate in a world starkly divided between those who have and those who have not, Crake is charged with designing new organisms to solve problems. These problems may be in the world—in terms, for example, of a dearth of organs for transplant—or in terms of the animals themselves—in terms, for example, of possessing undesirable characteristics that may make them unsuitable as companion animals for humans. The character Crake works with a design team on a number of new chimeras, including pigoons (pigs genetically engineered to create organs like kidneys compatible with humans, and carry more of them—e.g., six kidneys in a single animal) and rakunks (a new pet designed to eliminate the unsavory aspects of both raccoons and skunks into a new, adorable, domesticated companion). Religious and environmental cults, as well as other kinds of protest groups, organize against these kinds of genetic and biological manipulations, as well as the environmental degradation that such work either contributes to or merely papers over. Crake, in his philosophy and his work, appears to somehow straddle both sides. He sees humans as smart enough to be able to bend biological forces to our will, but not wise enough to benefit from such power. To that end, Crake bears primary responsibility for the disruption to the existing world—engineering a cleansing plague that destroys humans but leaves the rest of the biological world (including a few strategically engineered humanoids) relatively untouched.

Before that cataclysmic change, however, Crake and his team are tasked with creating a better version of the existing world. Part of that engineering task also and especially involves food, which informs our analysis in particular here. For Atwood, she wants to be sure that she includes scenes where her characters eat in her novels,[38] and food is a meaningful part of the trilogy. And what her characters eat in this collection of novels involves a kind of chicken and meat product engineered by Crake's team. The reader gets introduced to ChickieNobs, in particular, with a smaller mention of a product called NevRBled Shish-K-Buddies in the last book.

We first meet what become ChickieNobs as Jimmy is taking a tour of Crake's biodesign facility, cordoned off by several layers of biosecurity. The creatures Jimmy meets do not register—he has to be told that they are chickens, or chicken derivatives making specific parts (breasts or legs) to be precise. Modeled after a kind of hookworm, the creature is a protein-producing machine of the first order, with limited flaws (in terms of its main job—making meat). There are no extraneous features like eyes or a beak, making it difficult for Jimmy to orient himself (he does not even know where to look to "see" the creature). The efficiencies designed into the creatures extend to behaviors. Jimmy, still trying to wrap his head around the facets of the organism, engages in the following conversation:

"But what's it thinking?" said Jimmy.
The woman gave her jocular woodpecker yodel, and explained that they'd removed all the brain functions that had nothing to do with digestion, assimilation, and growth.
"It's sort of like a chicken hookworm," said Crake.
"No need for added growth hormones," said the woman, "the high growth rate's built in. You get chicken breasts in two weeks—that's a three-week improvement on the most efficient low-light, high-density chicken farming operation so far devised. And the animal-welfare freaks won't be able to say a word, because this thing feels no pain."[39]

Not the lab-based body of *in vitro* meat, but one simplified to the purest expression of protein the designers could muster.

Reactions to the product by consumers are mixed, though. Like the consumers of Chicken Little in *The Space Merchants*, those who eat ChickieNobs' Bucket o' Nubbins tend to be humans with fewer economic choices—a high-modern product that becomes the staple of the lower classes. Consumers of the Nubbins, either by choice or by necessity, become associated with baser desires and lower social standards. Jimmy, in fact, ends up eating these buckets frequently in his poorer days, much to the chagrin of his roommates at college, two of whom stop speaking to him as a result. Others confront the creatures in terms of existing schemas of animal-human relationships, treating these "protein machines" as extension of the chicken with which we may be more familiar. Though, for Atwood, this more traditional mode of interaction with these novel beings ends up in parody. When the plague commences, a few people use the ensuing chaos to try and release the biobeings into the world, to free them from their confines, including the ChickieNobs. They want to release these proto-chickens from their specialized environments and "return" them to nature—a place they were not designed for and never belonged. Newscasters report the scene:

> *Did you see that? Unbelievable! Ben, nobody can quite believe it. What we've just seen is a crazed mob of God's Gardeners, liberating a ChickieNobs production facility. Brad, this is hilarious, those ChickieNob things can't even walk! (Laughter.) Now, back to the studio.*

Many people do not know what to make of the protein or the organism, struggling to impose existing categories of understanding on a bespoke organism.[40]

Lab-based meat, more closely related to the contemporary *in vitro* meat efforts, also makes an appearance in the last book of the trilogy, *MaddAddam*. The book explores the lives of members of what might be termed a nature cult who have survived the plague and are refashioning their lives in the new world. Characters are not eating this meat in the book's present, but do have experiences from before, in the prelapsarian world. Those encounters describe a substance that is a noticeably poor substitute for meat derived from an animal. This lab meat has to be "camouflaged"[41] so you could not tell the difference. Zeke, a survivor, has another strategy: "NevRBled Shish-K-Buddies for those who want to eat meat without killing animals—the cubes were lab-cultured from cells ('No Animals Suffered'), and he figured that with enough beer they wouldn't taste too bad."[42] These characterizations have led Parry to point out what the author calls Atwood's nostalgia for meat, reflecting a desire for "real" meat rather than its lab-based or designed substitute.[43] Here, the world has both forms of meat, and *in vitro* varieties bear the stigma of inferiority—ironically, as chicken has been elevated to full "meat" status, its lab-based counterpart now comes to occupy the alternative meat category.

The world that Atwood creates focuses on and results from efforts to design our way out of complex problems. The products are there to replace any qualms one might have about how a problem was generated and still partake in the behavior that one is used to—to have one's cake and eat it, too. Worried about animal suffering, but really like chicken? ChickieNobs. Find raccoons precocious and skunks cute, but both a bit off-putting because of certain undesirable traits? Rakunks. An epidemic of renal failure in humans, and a dearth of qualified donors? Pigoons. Each of these might represent more complex social and ethical confrontations (e.g., whether or when meat should be eaten; whether any animal should be made a pet; what the underlying causes of kidney failure in humans might be), but are replaced by technical solutions. The age of design is resplendent with new products engineered to recalibrate the world to human behaviors, so that those behaviors are not so bad any more.

These solutions are not without their own problems, as one might expect, though the designers seem blinded to these possibilities, or at least enamored with their ability to control for them. Consider the following exchange between Jimmy and Crake, discussing a dog-wolf hybrid christened wolvogs and the possibility of that animal escaping into the wider world:

> "That would be a problem," said Crake. "But they won't get out. Nature is to zoos as God is to churches."
>
> "Meaning what?" said Jimmy. He wasn't paying close attention, he was worrying about the ChickieNobs and the wolvogs. Why is it he feels some line has been crossed, some boundary transgressed? How much is too much, how far is too far?
>
> "Those walls and bars are there for a reason," said Crake. "Not to keep us out, but to keep them in. Mankind needs barriers in both cases."
>
> "Them?"
>
> "Nature and God."
>
> "I thought you didn't believe in God," said Jimmy.
>
> "I don't believe in Nature either," said Crake. "Or not with a capital N."

Here, Crake expresses a sentiment about the need for boundaries, in a sense similar to the kinds of boundary drawing we explored in the first two chapters. There exists what is designed and what is not. There is order inside, and chaos out. To intertwine them both—in this case, the wolvogs and the wider world—would lead to devastation. Crake exhibits a common characteristic of fictional characters engaged in their own brand of world-making—hubris. He presumes that those barriers are impermeable. They are not. Atwood's post-plague landscape is rife with intermingling, as barriers collapse along with all of the other well-laid plans set out by Crake. What seems instructive here is this depiction of boundaries: the more critical the boundary, the more devastating the consequences of its inevitable collapse.[44]

Equally troubling in Atwood's novels, this emphasis on design ultimately gets directed at a new problem—humans themselves. More specifically, Crake shifts from attempting to design the world for humans to designing a world without them. Humans, Crake ultimately concludes, are fatally flawed in their nature—the technical solutions that he and his team have dreamed up have not made the world better, but just recalibrated human frailty and imperfections around different fault lines. People have not been improved by these technological breakthroughs, but rather have been given greater license to indulge their appetites. As such, the better tack would be to engineer a world without us. He does this through the development of two final products to help solve this last, most intractable problem. First, he designs a new humanoid species, dubbed by Jimmy as the Crakers. These humanoids are engineered by Crake to avoid what he sees as fundamental human flaws that lead to conflict and destruction (e.g., religious devotion, sexual jealousy, carnivorous diets), drawing certain biological features from other species. They have a digestive system borrowed from rabbits and a reproductive system borrowed in part from baboons. And the second product is a virus that is designed to kill all of humanity, save one (Jimmy) meant to shepherd the Crakers into their new world.

But like everything in Atwood's speculative fictional universe, none of these designs go quite to plan. Readers of subsequent books in the trilogy will know that Jimmy is not the only person to survive—many others managed to escape Crake's planned biological purge. As a result, the Crakers are at risk in their new Eden, not just from ill-intentioned humans but their own tendency to seek meaning and deification (in the guise of Jimmy, christened Snowman by the Crakers). The world always outstrips human intentions. This is one of the larger themes of the series, and represents a future that contains the desire to design the world to suit humanity, as well as the perils of actually doing so. Despite all of the power and control attached to this age of design, the arrogance of the orientation to the world always results in a kind of comeuppance or ironic turn. Technologies that desire total control ultimately prove inadequate in asserting it, with unexpected consequences. Importantly, those that plan the future cannot, by definition, plan for the unexpected. It is a concept that has a long history,[45] but this lack of novelty does not undermine the power of the message. Atwood places these designs in their larger context, where failure is expected and the consequences are dire. The walls between the inside and the outside are never foolproof.

There are lessons we might be able to glean from this exploration of possible protein-machine futures. There are various ways of imagining the future, and can involve both acts of creation (innovations meant to shape material relations) and acts of imagination (reflections meant to shape our thinking about what the future can and should bring). For the Chicken of Tomorrow contest, such creative acts certainly proved successful in generating a new world of chicken eating. The earlier effort of lab-based chicken culture, however, was more successful at spawning imagination than actually altering the material reality—what expanded was our fictional universe, not our concrete one. There is more real promise in contemporary efforts to culture protein outside of a conventional body, having generated both substance and hype for a new reality. And while it's not yet clear to what extent such efforts will move beyond boutique tinkering and into mainstream culture (will become, that is, the new chicken of tomorrow), such efforts have also spawned imagination and trepidation about the kind of future such creative possibility portends.

What both "ages" remind us is that, regardless of whether the material creation led to failure (or deception) or promise, or whether the fictional reaction was hopeful or cautious, the future will never scribe precisely in the manner anyone imagines. The Chicken of Tomorrow was successful at creating a new reality, but it was one that also reflected failures (as seen in the problems we explored in A Chicken, Part 1) and took unanticipated shapes. Thus, even as we look at the many possibilities of the future of meat, we should not assume that the future (when it finally arrives) will look like we imagine[46] or that it will be whatever we want it to be.

Some commonalities are worth considering for both ages, as they may help us think about the kind of future that might include *in vitro* meat technologies. Both ages (expansion and design) have to confront, redraw, and police boundaries and barriers—between new food organisms and the outside world, between the privileged and the impoverished. Those boundaries were bodies or were biologies that become labs or engineering techniques. What is inside and what is outside, what is included and what is excluded, become the province of human intention and human ingenuity. Those boundaries are physical (keeping unwanted substances out, and letting desired substances in) and they are ethical (insulating the consumption of protein from the moral challenge associated with the consumption of animals). And those boundaries include the ones drawn between classes of people—a stratified world that sees what to eat as a reflection of material circumstance (a choice for some and not for others). For those who are invested in the promise of the technology, a primary concern (reflected in PETA's contest rules) is price—who could afford to eat protein produced *in vitro*? To what extent will it remain the province of the rich and adventurous? In two of the works which imagine a world in which this technology has already achieved economies of scale, they both feature a world with clear differentiation of economic classes, and food operates as one factor of the dividing line. Those who eat the engineered food—the ChickieNob Bucket o' Nubbins, the NevRBled Shish-K-Buddies, the Chicken Little filets—are in the economically disadvantaged class. Those who eat this kind of protein are the ones without any choice. Those who are financially placed to be able to choose eat food produced more traditionally, more closely related to recognizable processes.[47] The highly engineered becomes equivalent with the cheap and the undesirable. The polarity shifts from its origin—the point closer to our contemporary world. *In vitro* meat is currently a very expensive and boutique item, but is imagined to shift to be common and cheap. And more conventionally grown food goes from common and inexpensive to rare and pricey.

Another commonality relates to the values of, and risks to, control over these processes. For boosters of the technology, being able to control the content and the process more fully, to produce clean protein no longer sullied by pathogens, unhealthy substances, and morally questionable actions. Those fictional worlds also explore the potential pitfalls of control, embodied in characters who feel they know how the world should operate, have the capacity to organize the world around those desires (at least initially), and make efforts to enact those changes. Courtenay does so toward his own life and his efforts to shape the desires of others. Crake does so toward his biobeings and his vision of a human-less utopia. Both also fall well short of the goal, with disastrous effects. Courtenay finds some remedy, siding with the Consies and reorienting his life to less consumption rather than more. For Crake, his actions are irredeemable, as he leaves a broken world for others to try and rebuild. Boundaries are drawn, and fall; designs are made, and fail.

These stories show us possible futures, while not exactly suggesting alternatives. If these potential worlds are undesirable, what kind of world might we strive for? The process involves not just learning lessons, but identifying the implicit fundamental questions that should be grappled with as we stumble into the future. Responsibility being fundamental. That will be our focus in the final chapter, setting up opportunities to stay rooted in the real, while still imagining a better tomorrow.

NOTES

1. To be fair, these elements of disruption were not planned. They were exactly the kinds of unanticipated realities that I had to accommodate, that made me realize that I had certain unacknowledged expectations.
2. Galusky, Wyatt. 2010. "Playing Chicken: Technologies of Domestication, Food, and Self." *Science as Culture* 19 (1): 15–35. As I detail here, I found that chicken raising was much more technically and philosophically complex than I realized, and my failures kept requiring me to reconfigure and reassess my goals.
3. See, for example, Latour, Bruno. 2002. "Morality and Technology: The End of the Means." *Theory, Culture, & Society* 19 (5/6): 247–60, and Verbeek, Peter Paul. 2005. *What Things Do: Philosophical Reflections on Technology, Agency, and Design.* University Park, PA: Pennsylvania State University Press.
4. Belasco, *Meals to Come*, 264.
5. Carrel, "ON THE PERMANENT LIFE OF TISSUES...", 528.
6. Carrel, "ON THE PERMANENT LIFE OF TISSUES...", 516.
7. A similar kind of thinking occurs in food systems—the logic of preventable occurrences gets applied to predation and bodily limitations on growth, which can be overcome through confinement and through reconfigurations of the body and its importance to the process. What humans want from the body (or the cell) become the most important consideration.
8. Carrel, "ON THE PERMANENT LIFE OF TISSUES...", 528. According to Landecker, important to the cultural resonance of Carrel's experiment and reporting related to the tendency of the heart tissue to pulse, thus manifesting its aliveness. Landecker, *Culturing Life*.

9. See Radiolab. 2007. *Life's Limit.* https://www.wnycstudios.org/story/91563-lifes-limit/.

10. Jiang, Lijing. 2012. "Alexis Carrel's Immortal Chick Heart Tissue Cultures (1912–1946) | The Embryo Project Encyclopedia." The Embryo Project Encyclopedia. July 3, 2012. https://embryo.asu.edu/pages/alexis-carrels-immortal-chick-heart-tissue-cultures-1912-1946.

11. Witkowski, J A. 1980. "Dr. Carrel's Immortal Cells." *Medical History* 24 (2): 129–42.—he favors the latter theory.

12. See, for example, Landecker, *Culturing Life*, 92: "the sense of control over biological matter was very appealing, and it was a period in which human control over biology seemed feasible and desirable. Tissue culture was used both as evidence of the human power to manipulate biology and as a demonstration of the dangers of unfettered reproduction. That is, it seemed to simultaneously represent biologists' ability to manipulate life and the potential anarchical powers of proliferation hidden in biological matter, which necessitated control. Scientists were more likely to see the death of their tissue cultures as failures of technique rather than challenges to the idea of life's indefinite bounds."

13. Notice the conception of what is absurd—growing parts of the animal that we do not want in order to get those that we do. Such practices are inefficient and undesirable (similar to what Edelman, et al., "In Vitro Cultured Meat Production," tout with *in vitro*) and potentially unnecessary—a mark of human progress (escaping the absurd) by getting what we want and avoiding what we do not.

14. Price, Jennifer. 2000. *Flight Maps: Adventures With Nature In Modern America.* New edition. New York: Basic Books.

15. He is especially worried about Communists.

16. This and subsequent quotes come from Oboler, Arch. n.d. *Lights Out (Old Time Radio).* http://archive.org/details/LightsOutoldTimeRadio. They represent transcriptions of the audio file composed by me.

17. Pohl, Frederik, and C.M. Kornbluth. 1969. *The Space Merchants.* New York: Walker, p. 76.

18. One might argue, at the risk of straining the metaphorical comparison past its breaking point, that the consies represent a kind of cancer on the society so envisioned, though in the logic of the novel, this would not be an apt comparison. The consies are seen as a positive force for change.

19. Landecker, *Culturing Life.*

20. Landecker, *Culturing Life*, 223–24.

21. See, for example, Kalof, Linda. 2007. *Looking at Animals in History.* London: Reaktion Books.

22. "New Harvest." n.d. New Harvest. Accessed June 8, 2018. https://www.new-harvest.org/.

23. "MOSAMEAT." n.d. Accessed June 8, 2018. https://mosameat.eu/index.html.

24. "Upside Foods." Accessed December 2, 2021. https://upsidefoods.com/. The original company, Memphis Meats, put it even more directly, striving "to bring delicious and healthy meat to your table by harvesting it from cells instead of animals. You can enjoy the meat you love today and feel good about how it's made because we strive to make it better for you...and for the world." See "Memphis Meats." n.d. Memphis Meats. Accessed June 8, 2018. http://www.memphismeats.com/.

25. "Tyson Ventures." n.d. Accessed June 8, 2018. https://www.tysonfoods.com/innovation/food-innovation/tyson-ventures.

26. "Beyond Meat – The Future of Protein™." n.d. Accessed June 8, 2018. http://beyondmeat.com/.

27. "Cultured Meat || Future Meat Technologies." n.d. Cultured Meat | Jerusalem | Future Meat Technologies. Accessed June 8, 2018. https://www.future-meat.com.

28. Ford, Matt. 2009. "In-Vitro Meat: Would Lab-Burgers Be Better for Us and the Planet?" 2009.

29. Zaraska, Marta. 2013. "Is Lab-Grown Meat Good for Us?" The Atlantic. August 19, 2013. https://www.theatlantic.com/health/archive/2013/08/is-lab-grown-meat-good-for-us/278778/.

30. Zaraska, Marta. 2016. "Lab-Grown Meat Is in Your Future, and It May Be Healthier than the Real Stuff." *Washington Post*, May 2, 2016, sec. Health & Science. https://www.washingtonpost.com/national/health-science/lab-grown-meat-is-in-your-future-and-it-may-be-healthier-than-the-real-stuff/2016/05/02/aa893f34-e630-11e5-a6f3-21ccdbc5f74e_story.html.

31. Gholipour, Bahar. 2017. "Lab-Grown Meat May Save a Lot More than Farm Animals' Lives." NBC News. April 6, 2017. https://www.nbcnews.com/mach/innovation/lab-grown-meat-may-save-lot-more-farm-animals-lives-n743091.

32. Javelosa, June. 2017. "Lab-Grown Meat Is Healthier. It's Cheaper. It's the Future." *Futurism* (blog). February 21, 2017. https://futurism.com/were-5-years-away-from-lab-grown-meat-hitting-store-shelves/.

33. Production systems would be reallocated.

34. Mattick, Carolyn, Amy Landis, and Brad Allenby. 2015. "The Problem With Making Meat in a Factory." *Slate*, September 28, 2015. http://www.slate.com/articles/technology/future_tense/2015/09/in_vitro_meat_probably_won_t_save_the_planet_yet.html.

35. Cannavò, Peter. 2010. "Listening to the 'Yuck Factor': Why In-Vitro Meat May Be Too Much to Digest." In. Washington, D.C.

36. Mattick, et al., "The Problem With Making Meat in a Factory," emphasis in original.
37. Atwood, Margaret. 2014. *MaddAddam*. Reprint edition. New York: Anchor, p. 393.
38. Atwood, Margaret. "Winton Tolles Lecture." Hamilton College, March 4, 2010.
39. Atwood, Margaret. 2003. *Oryx and Crake: A Novel*. New York: Doubleday, pp. 202–3.
40. Other novel biobeings fare much better in their new environments upon release, like the wolvogs and pigoons which become a new kind of predatory terror in this fallen world.
41. Atwood, *Oryx and Crake*, 61.
42. Atwood, *Oryx and Crake*, 233.
43. Parry, Jovian. 2009. "Oryx and Crake and the New Nostalgia for Meat." *Society and Animals* 17: 241–56.
44. Steinberg, Ted. 2006. *Acts of God: The Unnatural History of Natural Disaster in America*. 2nd edition. Oxford: Oxford University Press. See, also, Virilio's description of the shift from local to global problems, in Loeb, Zachary. "Inventing the Shipwreck." *Real Life*. January 3, 2022. https://reallifemag.com/inventing-the-shipwreck/.
45. As Jeff Goldblum's character famously put it in *Jurassic Park*, "Life, uh… finds a way." Spielberg, Stephen. n.d. *Jurassic Park (1993) – IMDb*. Accessed January 15, 2020. http://www.imdb.com/title/tt0107290/characters/nm0000156.
46. Wilson, Daniel H. 2007. *Where's My Jetpack?: A Guide to the Amazing Science Fiction Future That Never Arrived*. 1 edition. New York: Bloomsbury USA.
47. Parry, "Oryx and Crake and the New Nostalgia for Meat." See also Stephenson, *The Diamond Age*.

BIBLIOGRAPHY

Atwood, Margaret. *MaddAddam*. Reprint edition. New York: Anchor, 2014.
———. *Oryx and Crake*. New York: Doubleday, 2003.
———. "Winton Tolles Lecture." Hamilton College, March 4, 2010.
Belasco, Warren J. *Meals to Come: A History of the Future of Food*. Berkeley: University of California Press, 2006.
Beyond Meat. "Beyond Meat – The Future of Protein™," June 8, 2018. http://beyondmeat.com/.
Cannavò, Peter. "Listening to the 'Yuck Factor': Why In-Vitro Meat May Be Too Much to Digest." Washington, D.C., 2010.

Carrel, Alexis. "ON THE PERMANENT LIFE OF TISSUES OUTSIDE OF THE ORGANISM." *The Journal of Experimental Medicine* 15, no. 5 (May 1, 1912): 516–28.

Eagleton, Terry. *Reason, Faith, and Revolution: Reflections on the God Debate.* New Haven, CT: Yale University Press, 2010.

Edelman, P.E., D.C. McFarland, V.A. Mironov, and J.G. Matheny. "In Vitro Cultured Meat Production." NewHarvest.org, 2004. http://www.new-harvest.org/img/files/Invitro.pdf.

Ford, Matt. "In-Vitro Meat: Would Lab-Burgers Be Better for Us and the Planet?," 2009.

Future Meat Technologies. "Cultured Meat || Future Meat Technologies." Cultured Meat | Jerusalem | Future Meat Technologies, June 8, 2018. https://www.future-meat.com.

Galusky, Wyatt. "Playing Chicken: Technologies of Domestication, Food, and Self." *Science as Culture* 19, no. 1 (2010): 15–35.

Gholipour, Bahar. "Lab-Grown Meat May Save a Lot More than Farm Animals' Lives." NBC News, April 6, 2017. https://www.nbcnews.com/mach/innova-tion/lab-grown-meat-may-save-lot-more-farm-animals-lives-n743091.

Javelosa, June. "Lab-Grown Meat Is Healthier. It's Cheaper. It's the Future." *Futurism* (blog), February 21, 2017. https://futurism.com/were-5-years-away-from-lab-grown-meat-hitting-store-shelves/.

Jiang, Lijing. "Alexis Carrel's Immortal Chick Heart Tissue Cultures (1912–1946) | The Embryo Project Encyclopedia." The Embryo Project Encyclopedia, July 3, 2012. https://embryo.asu.edu/pages/alexis-carrels-immortal-chick-heart-tissue-cultures-1912-1946.

Kalof, Linda. *Looking at Animals in History.* London: Reaktion Books, 2007.

Landecker, Hannah. *Culturing Life: How Cells Became Technology.* Cambridge, MA: Harvard University Press, 2007.

Latour, Bruno. "Morality and Technology: The End of the Means." *Theory, Culture, & Society* 19, no. 5/6 (2002): 247–60.

Loeb, Zachary. "Inventing the Shipwreck." *Real Life.* January 3, 2022. https://reallifemag.com/inventing-the-shipwreck/.

Mattick, Carolyn, Amy Landis, and Brad Allenby. "The Problem With Making Meat in a Factory." *Slate*, September 28, 2015. http://www.slate.com/arti-cles/technology/future_tense/2015/09/in_vitro_meat_probably_won_t_save_the_planet_yet.html.

Memphis Meats. "Memphis Meats." Memphis Meats, June 8, 2018. http://www.memphismeats.com/.

MOSAMEAT. "MOSAMEAT," June 8, 2018. https://mosameat.eu/index.html.

New Harvest. "New Harvest," June 8, 2018. https://www.new-harvest.org/.

Oboler, Arch. *Lights Out (Old Time Radio)*, n.d. http://archive.org/details/LightsOutoldTimeRadio.

Parry, Jovian. "Oryx and Crake and the New Nostalgia for Meat." *Society and Animals* 17 (2009): 241–56.

Pohl, Frederik, and C.M. Kornbluth. *The Space Merchants*. New York: Walker, 1969.

Price, Jennifer. *Flight Maps: Adventures With Nature In Modern America*. New edition edition. New York: Basic Books, 2000.

Radiolab. *Life's Limit*, 2007. https://www.wnycstudios.org/story/91563-lifes-limit/.

Spielberg, Stephen. *Jurassic Park (1993) – IMDb*. Accessed January 15, 2020. http://www.imdb.com/title/tt0107290/characters/nm0000156.

Steinberg, Ted. *Acts of God: The Unnatural History of Natural Disaster in America*. 2 edition. Oxford: Oxford University Press, 2006.

Stephenson, Neal. *The Diamond Age: Or, a Young Lady's Illustrated Primer*. Reprint edition. New York: Spectra, 2000.

Tyson Foods. "Tyson Ventures," June 8, 2018. https://www.tysonfoods.com/innovation/food-innovation/tyson-ventures.

"Upside Foods." Accessed December 2, 2021. https://upsidefoods.com/.

Wilson, Daniel H. *Where's My Jetpack?: A Guide to the Amazing Science Fiction Future That Never Arrived*. 1 edition. New York: Bloomsbury USA, 2007.

Witkowski, J A. "Dr. Carrel's Immortal Cells." *Medical History* 24, no. 2 (April 1980): 129–42.

Zaraska, Marta. "Is Lab-Grown Meat Good for Us?" The Atlantic, August 19, 2013. https://www.theatlantic.com/health/archive/2013/08/is-lab-grown-meat-good-for-us/278778/.

———. "Lab-Grown Meat Is in Your Future, and It May Be Healthier than the Real Stuff." *Washington Post*, May 2, 2016, sec. Health & Science. https://www.washingtonpost.com/national/health-science/lab-grown-meat-is-in-your-future-and-it-may-be-healthier-than-the-real-stuff/2016/05/02/aa893f34-e630-11e5-a6f3-21ccdbc5f74e_story.html.

CHAPTER 6

Another Interval: A Fox, Two Chickens

Abstract This short section brings my own chickens back into the narrative, where their vulnerabilities highlight my responsibilities and how reliant I have been on luck.

Keywords Chicken • Backyard • Predation • Surprise

Being a hobbyist who lets his chickens have relatively free reign, I have also had to be alert to strange or stressful noises from the yard. It is how I discovered the two hawks who had taken the Buckeye, because they flew away as I was investigating the panicked calls of the two other birds who were cowering in the brambles. Other incidents I have been able to head off, either by luck or by alertness—heeding the distress of the chickens or the warning caws of crows. I caught a raccoon in the act in just that way, responding to a panicky kerfuffle. The predator had entered the coop, and snatched a young chicken. But I deprived it of its dinner; the raccoon only deprived the pullet of a few feathers. And there are a lot of false alarms,

W. Galusky, *Protein Machines, Technology, and the Nature of the Future*, https://doi.org/10.1007/978-3-031-08717-2_6

including chickens who are anxious, having just laid an egg and now not knowing where the other chickens are. Strangely, I have had two chickens just out of each other's line of sight exhibit cries that, to me, sound like the desire to find each other. Neither chicken makes a move, but just stands there and cries. Other chickens do not seem to be bothered. I may be wrong. Regardless, one certainly becomes much more sensitive to peril, taking a kind of chicken's eye view of the world.

It was mainly luck that resulted in the interruption of an attempted theft. On a snowy January afternoon, I happened to step out on the porch, investigating yet another strange noise in the yard. As my eyes readjusted to the bright snow, I noticed the Americauna near the garage, looking warily toward the front of the yard. Turning to follow her gaze, I saw a beautiful adult fox bounding through the snow, away from the house, with a chicken in its mouth. One of my chickens, as it happens—a Speckled Sussex (I call her Vashti). In panic myself now, I yelled, "Hey!" and again, "Hey!" Jacketless and shoeless, I was in no condition to launch into a chase (one I would lose regardless), but for some reason, the second hey prompted the fox to release the chicken and scamper away on its own. Lying inert in the snow was the Sussex.

Once I saw the fox had left, I went back into the house to put on appropriate attire, passing my spouse and telling her, "Vashti's dead." I am an eternal optimist. I walked out through the 8 inches of snow to retrieve her, past an array of feathers lost during the initial capture. As I reach Vashti, I notice her eyes are open, and I can hear her breathing. I pick her up—she seems to mostly be in a daze, and covered in snow. As I bring her back toward the house, I wipe the snow from her beak. She blinks, shakes her head, and promptly hops out of my arms, striding right back into the coop for the night. I check on her, but aside from some lost feathers, she seems no worse for wear.

There have been many of these moments in my years with the chickens (though few with this level of drama). They always refocus attention on that tension between liberty and safety, between enabling the birds to go where they please (even if it does not please me) and establishing a barrier between them and the dangers that would imperil them. We have three levels of confinement that come into play. There is a coop attached to a small enclosed run, which is the fullest level of security on offer. That is where they sleep and where they reluctantly stay in the morning until I get up and let them out. There is, in turn, a much larger fenced-in paddock area where they can roam more freely, but still have some limitations on how far. This is where they stay now when I am not at home, or when I have reason to worry about ground-based predators. The fence was mainly a compromise, as too much freedom enabled the chickens to not only wander on the road, but visit the neighbors' gardens and porches. The third level is, basically, the whole world, though primarily entails lounging under ferns and lilac bushes in the summer, browsing the yard, or perching on the deck furniture to groom. This is where they always prefer to be, and where I let them when I am home.

And this is where the four chickens were on an evening a few months later, in the spring, when I heard another weird chicken kerfuffle. I had been out with the chickens in the yard about ninety minutes earlier, and they had acted strangely then—kind of hiding in a copse of trees and expressing anxiety that I could not resolve and did not understand. When I went out on the porch later to see what was happening, I saw another or the same fox running the other way this time, in front of the porch, and carrying another chicken in its mouth—the Americauna, Buddy. I reacted much the same way as before, though not because I had some sense that it worked, but because it was the only thing that came to mind. Two "Hey"s in succession, prompting the fox yet again to drop the chicken to help secure its getaway. Upon being dropped, Buddy did not lay stunned but sprinted off to hide in the garage. I had trouble finding her to inspect her injuries—more severe than Vashti's (more lost feathers, a little blood), but I managed to clean her up and she is up and about, wandering around.

I know we were lucky. The timing was just right—another 15 seconds later in either case would have resulted in a much different outcome. I was also lucky that my missing the signals earlier that evening did not result in catastrophe (I now think they were alert to the fox or some sort of threat even then). But I still let them free, when I'm home, despite the risks. The next evening, the crows were cawing in the backyard, calling my attention to the fact that the fox was back, skulking around. Maybe I think my luck will hold.

CHAPTER 7

A Future, Part II

Abstract The final configuration speculates about the futures we have in front of us, and how the various forms of protein machines bring environmental and ethical stakes. Those stakes are especially pinned to the question of a sustainable future—one that is more resilient and not so dependent on control. The history of making chickens finds us seeking to maintain a kind of nature where animals are substances and humans are fully responsible (in creation and in maintenance) for the nature that keeps them alive. A more resilient approach to sustainability would include more, not less, proximity to animals and environments (not substances), sharing the burdens of responsibility by establishing relationships that need not be fully controlled to be functional.

Keywords Philosophy of technology • Inconsequentialism • Sustainability • Responsibility • Knowledge • Work

> ...for most of us, for decent people, the choice each day isn't between doing something good and doing something bad. It's between doing something good and doing nothing.
> —Jonathan Lee, *High Dive*, New York: Vintage Books, 2017

W. Galusky, *Protein Machines, Technology, and the Nature of the Future*, https://doi.org/10.1007/978-3-031-08717-2_7

In the novel by Hilary St. John Mandel, *Station Eleven*, the world has been engulfed in a flu pandemic that has decimated the existing human population, as well as the sustaining institutions (material and symbolic) that formed civilization. In this ravaged landscape, a troupe of performers, calling themselves the Traveling Symphony, caravan from community to community, putting on performances and plays (mostly Shakespeare) for the people who have survived. Despite the suspicion they face in each new community and the risks they encounter because of that suspicion, the members of the troupe promote a slogan remembered from a long-gone episode of *Star Trek*: "survival is insufficient." Cribbed from a piece of popular culture, used to motivate performances of what has become high culture, the slogan embodies a desire to inhabit a world not just to eke out some rough existence, but to thrive. The future they wish to build is not one of simple necessity, but of possibility and choice (even and in spite of the fact that things do not always go their way).

Mandel's story highlights how the future is partly a consequence of issues out of one's control (like a worldwide flu pandemic), but can still be influenced by the hopes and principles that guide one's actions. There are requirements that act as constraints on individual actions, and there are the possibilities that still exist within those constraints.[1] Our focus thus far has been on how people in the past attempted to configure their futures, and how those efforts led to our current situation. But we can also turn to discuss our own visions for the future, and the technologies that might help us get there. To do so, we need to be explicit with the questions we ask and the ideas we promote. By analyzing these past protein machines and how they have configured their future, and by looking at current protein machines and how they hope to do so, we can begin to think more critically about our own possible futures. We can ask if specific visions of who we are and what we should be able to do, embedded within these machines, are compatible with what we want for ourselves, understanding (as we do so) the various components that are implicated—animals, ecologies, people. I want to spend this chapter examining a theoretical approach to technology that emphasizes the contingency of technological design and that promotes the value and importance of understanding how technologies can shape our view of the world and ourselves.

The two protein machines we examined sought to exert control on the natural world, to heighten efficiencies and promote a specific output made desirable by the priorities of that system. The organization of the "machines" attempted to push to the margins all undesirable aspects of nonhuman nature that did not fit within the preferred outcomes. The traits that were marginalized include behavior, waste streams, and, as we moved into the laboratory, even skeletal structures and other aspects of the chicken physiology that humans did not eat or require. However, as we also explored, those marginalized elements did not simply disappear. They either appeared at inopportune moments as flaws that had to be accommodated or checked, or they had to be replaced by human-designed systems. We see those issues coterminous with our contemporary food systems; we should attempt to envision those type of issues emerging in current work on *in vitro* technologies. The works of fiction from the last chapter helped to demonstrate another sort of challenge—the orientation humans take toward the world and our machines, and one that assumes too much control and bears too much responsibility. Those works deal with what happens when our technologies outstrip our ability to direct and contain them. What will become important for us to see, as we explore an understanding of "technology" at the root of these relationships between humans, and nature, and non-human animals, is that these challenges are a normal and consistent part of the process. Technologies are designed in ways that render certain elements more visible and other elements less so, that produce disruption in how we understand and relate to the world, and that persistently test our capacity to manage them. These are fundamental characteristics of technologies.

This book has identified the food systems at the heart of our analysis as "protein machines" precisely in order to emphasize their technological aspects. As a result, we should seek to understand exactly what we might mean by technological. We should seek to make explicit the conception of technology we have in mind. For us, it should be one that reflects an interrelationship between humans and the natural world that is complex and conditional. That is, technologies help to create connections between ourselves and the world (including non-human animals) that are not strictly necessary—they could have been otherwise. This theoretical understanding will help serve as the basis for our earlier engagements with what chicken has become and how we might re-conceptualize our relationship to chicken in the future.

THINKING THROUGH TECHNOLOGY

Last chapter's reflections on the fictional depictions of chicken-less chicken created over the past 80 years can only take us so far, even as they help to emphasize the interconnectedness between protein machines like Chicken Little and ChickieNobs and the wider sociotechnical worlds that contain them. These narratives remind us about something important, that new configurations of technological systems do not occur in isolation. They are not simple additions that have no impact on their surroundings, but rather reflect a transformed world, and play a part in that transformation. Our explorations of the Chicken of Tomorrow contest showed us that, as well. Its ultimate success depended on not just a new chicken body, but a new production method and infrastructure, and a shifting vision of ourselves. So, while we participate in an era that is pursuing commercially viable *in vitro* meat production methods, as potential solutions to conventional protein production, we should at the same time consider to what extent we would welcome the types of transformations that will accompany them—in animals, in nature, and in ourselves. We need to ask, what kind of future do we want?

Real questions persist as to whether or to what extent this technology will develop into a viable meat production method, or whether it will be confined to the dustbins of history that contain other fantasies about future foods.[2] For many promoters and popularizers, the major uncertainties attached to this technology include three elements that are interrelated: technical capacity, cost, and acceptance. Technically, for example, issues involve creating relevant three-dimensional spaces for the protein to grow as an integrated muscle, and do so with the relevant "experiences" that will produce a substance that resembles muscle, and thus has the mouthfeel of meat. There are also the efforts to develop a non-animal-based serum for feeding the cells, in order to achieve the promise of animal-less meat. Connected to these concerns is the issue of price—to refine the technique at scale to enable the price to the consumer to be reasonable, comparable to other forms of protein production. A final uncertainty relates to acceptance—whether people would find the substance equivalent to existing protein to serve as a replacement, which is as much a question of taste as it is functional equivalence.[3] All of these elements add uncertainty to the process. They are also the most overt and anticipatory.

Another element clouding the contours of the future is precisely the unanticipated—the aspects of the technology as it becomes real (as it insinuates itself as a functioning machine within a transformed world) that are not expected in the design and implementation. These elements, what happens when an imagined design becomes a fact in the world, will be explored later, as we examine the characteristics of technology more generally. But it is important to understand that the levels of uncertainty are deep, and not fully in human control. Some issues are apparent, and may seem manageable, but others will be emergent and will need to be addressed at that specific point in time. Uncertainty abounds.

Despite this uncertainty, and even because of it, we can still offer reflections on a possible future compatible with *in vitro* meat technologies. We can do so by exploring the assumptions connected to that technology— assumptions about the world, about animals, and about ourselves. That is, if we can develop a fuller understanding of what the stakes are and what kind of transformations might occur, we can at least grapple with a kind of future made possible by *in vitro* meat technologies and whether that may be the kind of future we want to pursue. To what extent are these protein machines compatible with our visions of the future, and our place in it? What kinds of responsibilities will that technology create, and are we ready for them? We may not be able to identify the specific complexities of a specific future, but we should be able to sketch the contours of the most basic transformations.

Conjuring up a view of a possible, and desirable, future can be difficult. Much easier is the dystopic vision—what we don't want. The fictional narratives we explored mostly look to an already formed world—one that best serves as a warning about what could go wrong. They provide a fully realized environment in which protein machines play a significant and consistent, but ultimately small, part in a fallen world. These narratives provide a kind of warning, but do not provide a guide for today nor for the near, plannable future, as a way to chart a new and better path.[4] Kathryn Schulz, writing about potential west coast earthquakes in *The New Yorker*, puts the challenge this way: "we excel at imagining future scenarios, including awful ones. But such apocalyptic visions are a form of escapism, not a moral summons, and still less a plan of action. Where we stumble is in conjuring up grim futures in a way that helps to avert them."[5] How might we think about our shared future not as inevitable, but as something fungible and worth trying to shape? How might we understand the roles technologies can play in shaping that future? What kinds of relationships to the natural world and to ourselves do technologies facilitate? And how might specific protein machines participate in those relationships? Thinking through those questions will allow us to confront the kind of future we may want to pursue. The goal here is not to define what the future should be, as that will ultimately be a collective project. Rather, the chapter sets out to sketch the means by which we might be able to do so, thinking through technology.

THE FUTURE AS A COLLECTIVE PROJECT

Of course, we could just think of the future as something that happens to us, or at best a set of prescribed options from which we are empowered to choose. Genetically modified foods are often conceptualized and experienced in this manner, as something we might choose to buy or not buy. We might even make it our mission as consumers to agitate for that form of market-based choice—lobbying for policies that require labelling of products based on their connections to certain biotechnical processes (e.g., genetic modification, or cloning).[6] Placing action at the level of the consumer, though, represents a very impoverished form of voice, reducing individuals to providing market-based opinions connected to purchasing power on a state of affairs that already exists. We can decide to purchase these products or not, all things being equal,[7] but we can't decide to live in a world where these products do not exist. These kind of actions at the level of the individual connect to what Vogel refers to as inconsequentialism.[8] This phenomenon, which we will explore more below, identifies this kind of individualized activity as ineffectual in shaping the future, given its limited impact. Worse, such a decision to make a consumer choice based on one's values may actually be irrational—why spend more in the marketplace to purchase such items, if such a purchase will have no discernable difference? It would be more rational to maintain one's relative economic position by pursuing goods independent of values aside from price.

Our focus, instead, will be on conceptualizing a future as an act of collective responsibility—not just as something that happens to us, but as something that we can collaborate on choosing and help shape. That sense of responsibility is important because it connects to the role that technology plays in shaping the world and to the roles that humans can take on in facilitating that shaping. Situated in its historical context, the story of the Chicken of Tomorrow looks more contingent, not the inevitable result of an inexorable march toward larger, more consistent chickens and greater numbers of chicken eaters. It's possible that chicken could have been consigned to more marginal "meats" in the United States. That a group of people sought a different future and brought it into existence reminds us that our world has been and can be shaped by human intention—that the world can be made and remade based on decisions we make. The goal for us here will be to take that contingency seriously, and widen the scope of possible input. To claim as a good the broader impact of society in the shape of the future. As Preston notes, discussing another form of futurescaping, synthetic biology:

> Stakeholders should be given the opportunity to know what is coming their way and offered meaningful input on whether they actually want that type of future. If the extent of this input is simply an after-the-fact choice about whether to purchase a particular end-product, then too much has already been decided. Too little has been shared about what is happening to their world. […] the future of nature should not be determined simply by what is possible. *Can* had never automatically entailed *should*. The shape of the future must involve deliberation and discussion by as many people as possible.[9]

We should take Preston's injunction seriously, and take a more active role in configuring our future. To do so, we need to be able to consider what kind of future we want to pursue, and to consider the kinds of protein machine configurations that are compatible with that pursuit. We need, in other words, to reflect on just how technologies enable us to achieve our intentions. Here, then, I want to turn to a brief discussion about technology as a means of engaging and transforming the world, before applying that analysis explicitly to our protein machines and the choices that we should confront.

TECHNOLOGICAL CHANGE

Technologies change the world. They represent the inscription of human intention onto nature, a way for us to shape the world into something we want. That's part of the reason we can discuss possible futures and the technologies and natures that might populate them. Critically, however, those intentions are inscribed onto nature imperfectly—there are limits to how completely and how consistently we can change the world to reflect our designs. The limitations come from a much fuzzier distinction between humans and nature than we might tend to assume, and from an intractable wildness in the nature that we attempt to tame. To explore the full scope of these arguments is beyond the scope of this project, but I want to offer a quick overview of this imperfect translation between human design, technologies, and nature, because it speaks to a crucial sense of responsibility we have to the worlds we attempt to create, including those that contain different forms of protein machine.

Vogel, in his book *Thinking like a Mall*, collapses the easy distinction we might draw between nature and technology. His argument is complex and nuanced, and takes pains both to trouble that distinction and, at the same time, to retain some wildness in nature. It's this latter point that is the most important for us here. Collapsing the boundary between nature and technology can suggest that we can simply design the world to reflect what we want that world to be. We can create protein machines that do our bidding. For Vogel, however, nature does not become the *mere* product of human activity. Rather, our actions interact with and shape but do not determine what we come to know as nature. He argues that human intention, or at least the capacity of humans to assert their intention in the world, is limited: "We construct the world through our practices, not through our ideas. Practices are not 'ideal' but rather are entirely *real*. Engaging in them takes work, meets (indeed, requires) resistance, and frequently leads to failure."[10] Moving intention into the world requires confrontation with a reality that is different than us and will not simply comply.

Take, for example, working with wood. One might have in mind the perfect bookcase. Translating that ideal into something concrete is a tricky business. I know this from experience—I can never seem to get my hands and the wood and the tools to work in harmony so that I produce precisely what I set out to. I can get close, and do get closer with more practice and better technique, but never perfect. The *real*, as Vogel conceptualizes it, is always putting paid the idea that we can simply do what we want in the world: "The real is independent (unlike the ideal, or the thought, or the intention): that's what reality *is*."[11] We have to work with the natural world, and compromise our ideas to what that world will allow. Just as our own bodies are not simply at our full command. Nor are chicken bodies, which as we saw in the earlier chapters resist attempts to control and direct them fully toward human desire. They maintain a reality that is not subsumed and must ultimately be reckoned with. The presence of the real and the recognition that nature is partly the result of human intention reemphasizes the moral contours of these interactions—how best do we engage with and attempt to transform nature, knowing that nature must agree to the transformation?

Technology, then, is not apart from nature, but within it, and ultimately involves imperfect attempts to translate human intention into real circumstances. To describe this translation as imperfect does not imply that it is unsuccessful. The process of creating a chicken of tomorrow—this sprawling, conventional protein machine we explored in the first section—was certainly a kind of success. Such a version of the chicken was created and transformed into a real entity in the world. The imperfection—the fact that human intention cannot be completely inscribed in the world—is why the problems also continue to persist, like those that adhere to conventional food production systems. This distance between what we want and what we get also suggests reasons why we might seek a better way to configure the protein machine, where we won't eliminate problems, but rather encounter different kinds of problems that are more compatible with our shared values.[12] Such a concept of the interplay between technology and nature also helps to describe the current state of *in vitro*, struggling to make real the possibilities inherent in the technique—to turn intention into reality. The issue is not just about what form such a product will take as a real thing, but also what kind of problems might emerge associated with that translation.

These imperfect successes, like the Chicken of Tomorrow, must also be understood not as singular moments, but as sustained endeavors. It's not enough to create a protein machine—one must also generate and maintain it. This temporal element adds another dimension to the analysis. We must seek to create a future and sustain it. We must, in other words, be capable of identifying and fulfilling the responsibilities that are required by these technologies and natures we strive to build. To create and to maintain—two key elements connected to the idea of responsibility as world builders. And it is toward this set of responsibilities we would bear that we now turn.

Technologies and/as Responsibilities

The two key ways of understanding responsibility for technologies (and, for Vogel, nature itself) come in two forms: causal and moral. In the causal sense, we have brought into being specific forms of things that would not otherwise exist. We are the agents of causation, what Heidegger, borrowing from Aristotle, refers to as the efficient cause—the individual who brings together materials, form, and context to create a thing.[13] In our particular frame of reference, humans brought specific forms of chickens into existence, creating a kind of being that would not have existed as such without our intervention. Importantly, following our discussion above, causal responsibility is not the result of magic. We have to work with real materials and real existences other than us that may resist, and not all creations succeed. Humans are one important reason for what does succeed, but this is a process of negotiation with skill, material, form, and context.

Responsibility is not limited to creation, however. We also bear some burden for caring for that creation—for what happens to those things in the world that now exist because of us. At least, as Vogel notes, in relationship to the natural world we help to shape, "[we] ought to be responsible for them in the sense that it should *matter* to us what happens to them— how they are used, how they are destroyed, and, most importantly, simply what they are. *We are responsible for the environment.*"[14] It is not just the fact that we should recognize our causal responsibility for creation; we should also recognize our moral responsibility for the life of that thing— how it is used, tweaked, or terminated. How it thrives or struggles. Latour makes a similar claim, reminding us of a key lesson from Shelley's *Frankenstein* (not that Dr. Frankenstein created the monster, but that he subsequently abandoned it) can be applied to our dealings with technologies:

> It is not the case that we have failed to care for Creation, but that we have failed to care for our technological creations. We confuse the monster for its creator and blame our sins against Nature upon our creations. But our sin is not that we created technologies but that we failed to love and care for them. It is as if we decided that we were unable to follow through with the education of our children.[15]

Having decided to create something in the world, through that negotiation between intention and reality, we carry the burden of that creation. We have to care for it, because it would not exist as such without us. And we have to care for the world that it helps to generate. Technologies entail responsibility—for causation and for care.

I want to extend our understanding of responsibility beyond those twin relationships, to include *how* those relationships are established—through what mechanisms. That is, what allows us to generate new things in the world, and what is required to sustain them? For this, we can look to the practical elements of responsibility, knowledge and work. These are aspects that add their own layers of complexity in terms of thinking through our technologically mediated worlds.

Let's take the first element, knowledge. In order to act in the world, we need to have some knowledge of how that world works. At least, we have to know *enough*. In order to get a plant to grow in my garden, for example, I have to at least be aware of the need for dirt and water and seeds. Also sunlight. I don't have to know much about seed biology for this level of engagement. Like Thoreau, I may just need to have faith in a seed.[16] But at this level of intention, I can have success. Of course, to get weeds to grow, I need know even less. They grow of their own volition, rather than mine, which is why we refer to them as weeds. Though I ignore weeds at my own peril, because they can prohibit the growth of what I am actually after.[17] As a result, I may need to know how to check their growth, which often involves just identifying them and pulling them up. And the more I want my garden to look like I imagine it, with robust plants laden with vegetables, the more I will need to know—about plant growth and nutrition, about pest and disease and weed management. I can get lucky, of course. And I have—the first season I planted a garden, I had a confidence boosting honeymoon period. This gardening thing is easy, I thought. But as I gardened in subsequent seasons, more pests showed up. Less optimal growing conditions reared their blight-ridden heads. And my lack of knowledge was exposed, as garden after garden failed to produce the results of that first blessed season.[18]

Knowledge does not guarantee success, of course, though it helps. One key aspect to recognize here is the way in which knowledge operates on a sliding scale. The more we know about biology, for example, the more we can intervene in the development and composition of seeds. If we want to modify the genes within those seeds and more precisely produce a desired product, faith will no longer do.[19] Knowledge is required. I do not want to get into an extraordinary nuanced discussion of knowledge here—as to what precisely it means and what does and does not count. Importantly, for our purposes, knowledge is meant to signify what is required of humans in order to get the technology to function somewhere close to how we intended—how intention is translated into the function of a thing. We need to know what to do, and what not to do; we need to know what to allow, and what to forbid. We are responsible for knowing enough to enable some level of success.

This kind of requirement can be seen throughout the development of our protein machines from the previous chapters. Knowledge was required to turn the chicken of that day into the chicken of tomorrow, to render a bird more fully a part of the larger protein machine configured to maximize protein production: breeding, nutrition, physiology, behavior, reproduction. To intervene specifically and not haphazardly requires knowledge systems capable of achieving the desired outcome. To produce the correct breed of bird, to manage the efficiency and consistency of its growth, and to maintain its health under these conditions. Each form of intervention requires fuller and more accurate knowledge about the chicken as such. The same can be said for *in vitro*-based techniques, especially in the almost heroic act of recreating a body in which muscle can be grown and developed. Within the *in vivo* chicken, the body operates as a cultivator of muscle, at least in part. And that function operates relatively independently. I don't have to know how that body produces muscle to benefit from it doing so. But the body has other demands that may run counter to our ultimate aim of producing protein, including (as we noted in A Chicken, Part 2) feathers and bones. So, what if we could do without the chicken to produce chicken? *In vitro*, in removing the animal body, has to recreate the essential functions the body plays in protein production. But what is essential? And how can it be simulated? Some key aspects appear to be motion through space and providing a well-regulated environment for growth (providing nutrients and protecting from disease).

At least, those are some of the main functions we recognize and thus attempt to replicate, by stimulating the muscle tissue mechanically or electrically, providing nutrients though either regulating cell thickness or building scaffolds, and by maintaining a hyper-sterile environment. The point of emphasis is this—*in vitro* techniques enable a much more extensive application of our intention, but require much more complete forms of knowledge. This requirement has grown in the transition from the Chicken of Tomorrow, which sought a specific kind of chicken body, to the potential chicken of the future, which seeks to dispense with that body and replace it within a lab-based context. Knowledge is one cost for this level of control, as the margin for error is smaller, with no existing body to help cultivate the cells we seek. Thus, it is important to note that knowledge is required for all types of technological interaction. But that requirement is steeper in some forms of technological engagement. For the chickens living in my yard, I can know surprisingly little. For the industrialized chicken of tomorrow, much more knowledge is required. For the laboratory-based chicken of the future, more still.

Even as we come to recognize that greater intervention requires greater knowledge, however, we shouldn't assume that such acquisition occurs in a straight line. Our own intentions are modified in the process of achieving our goals, so that our goals themselves can be modified. Let's take the first step of that modification in the context of more generic engagement with technology. This change can be challenging to notice because of how easily we can accommodate ourselves to tasks. If I were to want to hammer a nail, for example. This action appears relatively easy, especially if such activity is practiced. But we shouldn't overlook how I had to learn to grip the hammer, and to swing it, and aim it. I had to learn how to hold the nail without putting any of my fingers in peril (something learned, and relearned, the hard way—it's a skill that also involves aim). I had to learn the properties of material meant to take the nail—hardness and brittleness being particularly important qualities. These were abilities or knowledge I had to pick up on the way to my destination, as a means of imposing my intention. Interactions with tools, and methods, and materials.

Bruno Latour calls this type of modification a detour.[20] The concept conveys the sense of delay or deflection that occurs as we seek our technologically mediated destination—all the little stops between the point A of our designs and the point B of the actual outcome. Such detours can be seen in our animal-based explorations in this book. Take the decision I made to get chickens again. That was the goal I had, which suddenly necessitated a host of other decisions along the way, shaping not just the chickens I wanted to get, but how many. For example, take the coop. I had to consult plans, make decisions, learn how to cut boards on an angle to create a pitched roof, learn what kind of roosts to build to avoid frostbitten toes in winter. Skills for joinery and for sizing and cutting boards. For nailing barn siding. For laying linoleum tiles. Trips to the hardware store and the salvage yard. Detours taken on the way from getting the notion of raising chickens to having the capacity to *actually* have them. The same can be said for the desire to make chicken *in vitro*, without a body. Those needs, to replace vital functions in the creation of muscle, based on techniques and limitations in materials, are also detours between the desire to make muscle, and the arrival at edible meat protein. Techniques that need to be learned, which also shape what the real output is, rather than the ideal output.

For Latour, the concept of detour does not just apply to diversions in the process, but in the destination itself. Our initial goal is often adjusted slightly, or profoundly, based on the turns we take in the encounter with the world. My ultimate goal to hammer a nail may be delayed because of injury, moved because of an unsuitable location, or abandoned in favor of a different fastener due to material conditions. My experience with back-yard chickens had gone through many iterations—from a radical free-range advocate, to the erection of a temporary and then permanent fence to placate the neighbors and keep the birds relatively safer (though the more agile birds promptly flew over it), and finally to the installation of netting to protect the bantam chickens from a persistent Cooper's hawk. My well-designed nesting boxes, with outside access, did not stop me from finding eggs in the most random places.

Each element of those detours suggests that, even though knowledge may be a necessary element of intention, it is not a simply translation of idea into reality, because the material world and our movement in it require some modification. As we seek to modify the world through the mediation of some technological system, we in turn are modified by it. This kind of change is required because we are moving from ideas about how we want the world to be, to create something that exists in the world using tools that make their own demands. Latour notes that we may be reluctant to see the ways in which this interaction alters our initial goals, but:

> If we fail to recognize how much the use of a technique, however simple, has displaced, translated, modified, or inflected the initial intention, it is simply because we have *changed the end in changing the means*, and because, through a slipping of the will, we have begun to wish something quite else from what we at first desired.[21]

So knowledge as responsibility, in configuring this part of our relationship to the natural world, is not simply about imposing our will on that world, but also having to adapt ourselves—in terms of learning/doing new things, and in terms of adjusting our goals. Why it matters that we confront the future not as merely the present plus a new/substitute technology, but as a newly realized world. This element of technological engagement, of machine building, suggests that we do need to consider how we change with technology (even as we design and change technology to suit ourselves). This type of analysis should remind us of the lessons from A Future, Part 1, at least in as much as it is important to imagine not just a new technology, but a new world in which it fits, and the ways in which we may have changed along the way. A speculative enterprise, to be sure.

There is a second aspect of this component of intentional engagement of the natural world. It's not just that knowledge does not allow us to follow a straight line between our initial designs and the ultimate outcome. It's not just a question of detours. It's also a question of imperfection or incompleteness. When tasked with translating our intentions into something actual in the world—whether a driven nail, a more predictably protein-packed chicken body, or a clump of cells unsupported by an animal—we can build up robust systems of knowledge based on the questions we think to ask and the feedback we recognize. But we have a long history of confronting, only later, the unanticipated consequences of our work. The split piece of wood or bent nail, the overstressed skeleton of the newly muscle-laden animal. Human history is littered with examples of interventions that outstripped their original intent, especially true in the introduction of novel species for the purposes of erosion control (kudzu), the use of additives for internal combustion engines (ethyl lead), and the creation of medicines for the treatment of morning sickness (thalidomide). Each of these outcomes were unanticipated and undesired.

Of course, just because an outcome is unanticipated does not mean it has to be negative. Examples can also be found to demonstrate positive transformations that resulted from accident. But that's part of the point. Actions in the world do not simply adhere to their explicit design. Just as not all outcomes need be noticed or knowable. There seems to be good reason to believe that there will always be a gap between what we know the world as, and the world itself. Adorno called this gap the "remainder" between the object in the world and the concepts we use to understand it.[22] We can shrink the remainder, but not get rid of it. Vogel offers another way of thinking about it: "There is a *gap*, in the construction of every artifact, between the intention with which the builder acts and the consequences of her acts, a gap that is ineliminable and indeed constitutive of what it is to construct something."[23] The incompleteness of knowledge systems does seem likely to be a fundamental aspect of the human conceptual apparatus. But even if it were simply a common type of error, we could still confidently argue that the likelihood that new technological systems will be imperfect or incomplete and lead to unanticipated effects is very high. What this reflective analysis shows is that one requirement for asserting intention in the world is developing knowledge of that world. This responsibility is not neutral, as in seeking to develop and apply that knowledge, we modify ourselves and the world, and do so with systems that, despite their power, are still imperfect. All of this reflection on knowledge and its diversions and imperfections is not to denigrate the technological achievements that we have brought about. Rather, it is meant to begin to sketch its limitations as we consider the question of our own shared future.

Alluded to in our discussion above is another requirement of responsibility coupled with knowledge—work. It's not just knowing *how* to do something, but it also involves actually doing it. Or, at least, enabling it to happen. For this element, the hammer may not be the most illustrative example, because the work seems relatively minimal. But it's certainly true that if I want to get the hammer to swing through space in a direction compatible with my designs (to strike, say, the head of a nail), I have to do the work of the swinging and the aiming. I work in concert with gravity and momentum, but perform a vital role in the successful use of the tool.[24] Perhaps a more complete example of the work required to impose our intention on the natural world can be found in John McPhee's essay, "Atchafalaya."[25] In it, he details the herculean efforts by the US Corps of Engineers to keep the Mississippi River from being captured by the Atchafalaya River, a body of water that offers a more direct route to the Gulf of Mexico than the one taken by the Mississippi currently.[26] Left alone, and adhering to natural imperatives, the Mississippi would seek a shorter route to the Gulf, and thus bypass much of southeast Louisiana, including New Orleans. This expedience would be catastrophic to the economic and industrial wellbeing of those human settlements currently downriver from this capture point. So, the Corps worked to prevent that from happening, to impose its will on the river, and installed a river flow control mechanism that kept the Mississippi on its desired path. To do so, the Corps had to understand enough about the flow and forces of water, as we explored above, but also had to expend lots of effort—energy needed to install, regulate, and repair the control mechanism and surveil the Mississippi river, lest it overwhelm the mechanism. Because we wanted the Mississippi to stick to its current course, because so much of our economic infrastructure depended on that course, we worked to ensure that it stayed that way. Managing the flow requires an enormous effort to keep the river in the place we want it to be.

Less effort is involved in caring for backyard chickens. But consider mine, at any rate. As I took those various detours and modified what I came to understand as "free range," I kept committing to more labor. I was already aware of the basic agreement—getting up to let them out in the mornings and securing their coop in the evenings. I even built an enclosed run to give them more secure space and to give me more guilt-free time in the morning before letting them out. But during the day, things changed. To keep my chickens in a specific space, I first relied on a kind of timidity in the birds, assuming that they would not want to wander too far from their home. When that failed, after they became brave enough to venture too far,[27] I built a kind of makeshift fencing. Later, in deference to aesthetics and effectiveness, I had a split rail fence built to create a paddock area. Later still, because the fence failed to stop my more nimble and brave birds from jumping up and over it,[28] and because hawks found it no barrier at all, I ultimately installed netting. Because they range less widely now, I need to supplement their diet more—more frequent topping off of the feeder, more frequent trips to the feed store. It's not labor I begrudge, but labor I acknowledge as such. I have done and continue to do more for my chickens, so as to exert slightly more control over their lives.

Scaling up into the realm of the Chicken of Tomorrow, we see an even greater translation of effort, away from the animals and toward the humans who have specific designs for them. The chickens are selectively bred, which means that people are now overseeing the selection and mechanics of reproduction (fertilization and incubation). This is work now done by humans, or arranged by them through systems of electricity and engineering. Chicken growth is centralized, so people are transporting chicks to a designated space to grow them to their desired state. Chicken diet is engineered and regulated, so people grow and process food, distribute that food to grow houses, and feed the chickens. Chickens are largely kept indoors, so people have to design and distribute supplements to replace the sun.

There is of course more labor involved in the death and consumption of the chicken, even more so when the chicken meat is further processed into engineered shapes. But by focusing on the new life of the chicken, in a way that reflects more ambitious human intentions, we can see that activities chickens used to do for themselves are now done by humans. Chickens don't feed. They are fed. They don't move. They are moved. They don't breed. They are bred. The passive construction contains the shift in burden, from the animals to the humans who direct their existences more fully. Certain behaviors and activities still remain with the animals that aren't captured in the lifecycle here. Chickens will still seek to dust bathe if possible, or clean themselves. They will also seek to establish a pecking order, which may prompt even more work—not to replace the behavior but to stop it.[29]

The move to replace the body of the chicken with a lab-based substitute comes with a further translation of effort from chicken to people. We must design and implement techniques that replace movement, that replace vascular function, that replace regulatory function, that protect from disease. We have to do the work of the body, in order to supplant it. The consequence of our efforts to further impose our designs on the form of the chicken come at the cost of doing more work on behalf of that form. And it is not strictly human labor that does this work. Certainly, people are involved in performing specific tasks, like those who produce the feed or harvest cells. Some of the labor is dedicated toward organization and logistics, arranging when things happen and to what extent. And some of that effort, perhaps the lion's share, is found in directing mechanization toward these newly necessary tasks. The farm implements (themselves engineered and built and shipped) running on petroleum (extracted, refined, and shipped), moving, feeding, and cleaning up after the animals. The lab equipment powered by electricity used to stimulate the cells to simulate movement.

The process of creating objects that reflect human designs has always taken work. That's the role that humans take on, and what makes us one of the causes of that thing, because we have the capacity to organize elements toward a specific goal. Here's John Locke reckoning up the value of a loaf of bread:

> ...for it is not barely the plough-man's pains, the reaper's and thresher's toil, and the baker's sweat, is to be counted into the bread we eat; the labour of those who broke the oxen, who digged and wrought the iron and stones, who felled and framed the timber employed about the plough, mill, oven, or any other utensils, which are a vast number, requisite to this corn, from its being feed to be sown to its being made bread, must all be charged on the account of labour, and received as an effect of that[.][30]

All of that work, just to produce a loaf of bread. Locke, writing in the seventeenth century, shows awareness of the amount of energy required to translate nature into something that we want, by mixing our labor with the land. But while such awareness is not novel, it is also not common. Often, when we encounter a loaf of bread, pre-sliced, we do so as a convenience, something that has made our lives easier. And we fail to reckon up the labors that went into its production. We don't recognize that effort hasn't decreased as much as shifted, sometimes from animals to humans, from natural systems to human systems, from humans to machines. Regardless, it is a translation of effort that becomes the responsibility of the humans who seek to orchestrate it. This kind of awareness is part of what Wendell Berry advocates for when he reminds us that "eating is an agricultural act"[31] and what causes Michael Pollan to christen a modern steer as "another fossil-fuel machine."[32] We will want to remember this kind of labor-related necessity when we consider our task of projecting our future.

To recap, we have explored the requirements that accompany our desires to insert our intentions into the world—to make the world look the way we want, behave the way we expect, and produce what we want to possess. These goals require us to cause and to care for the world we help create, but only to the extent that we are also able to know and willing to do what is necessary. Each of those responsibilities is not simple. They come with their own complexities. But they also generally reflect a translation of responsibility from natural systems to human ones. Importantly, the translation of these responsibilities is not strictly linear. We do not always seek to assert greater intention on the world. Sometimes, we might choose less control over natural systems by letting them grow wild, or letting them be more susceptible to risk. But that's the point. These forms of engagement with the world are contingent—they are a matter of choice. And as we think about the choices we make about the kind of world we might want to inhabit, we need to be aware of the stakes.

There is one more element of this translation that is worth exploring at this juncture, which involves the roles people play within these technological systems. As we design and create technologies that more fully express our intentions, and as we take on more responsibilities to make those objects real, we may also, paradoxically, involve fewer humans in their production. I have never made a hammer, though I have used plenty. I doubt I am alone in that fact.[33] Contemporary hammers are relatively cheap, relatively consistent, and relatively effective once we correct for user error. My connection to the hammer is strictly as a user. It would not even occur to me to attempt to construct a permanent replacement, and replicate the forged head, the plastic handle, and the rubber grip (even that rudimentary list of components I hadn't really considered until doing this mental exercise). A lot of knowledge and effort by others went into its design and production, and lets me focus on other things. I could attempt to create my own, but such an effort would likely lead to more than a few absurdities. Consider, for example, Thomas Thwaites, who thought it would be a good idea to build a plastic toaster from scratch, taking 9 months and costing 250 times what one might cost in the marketplace.[34]

The important point to make here is that the complexities associated with convenient technologies often lead to more users of a technology, and fewer producers. And if we follow the trajectory of increased knowledge and effort set out above, the results occur because fewer people are necessary and fewer people are qualified. The latter point is perhaps the more intuitive. Certainly, as the type and depth of knowledge increases, fewer individuals will be able to offer that knowledge. The capacity to grow cells *in vitro* requires an elaborate and specialized form of knowledge possessed at least by the people directing the process, if not the technicians executing particular tasks. Were I to be confronted with a lump of cells or an *in vitro* based meat product, I would have a difficult time even imagining how it was produced, much less being able to do so myself. In fact, to try to do so would require a substantial amount of investment in both time and capital (and one of the reasons why Thwaite subtitles his quest to build a toaster as "A Heroic Attempt"). It's easier to imagine raising a chicken, though the scale and precision of an industrial setting, especially to produce a competitively priced animal, is particularly forbidding. I could certainly buy the engineered feed and the selectively bred chicks, but would be challenged to replicate the scale and precision of manufacture. Much like what happened within the shift toward industrialized methods, growers become more and more the executors of the will of the poultry companies—doing what they are told, when, at the risk of losing their contracts for failure to comply.[35] Fewer people, even in this form of organization, had the capacity to direct this newly engineered chicken body, environment, and production processes. As a hobbyist with no real economic imperative, I can participate in the raising of chickens, but would blanche at the prospect of having to do so competitively.[36]

These moves also required fewer people to be a part of the process, even as more work overall was being applied. As we noted above, the amount of work is not just direct human labor, but machine labor directed by humans. Chickens were no longer required to find food and feed themselves. Now humans did it for them, by organizing the labor of other humans and of machines consistent with the industrialization of food generally.[37] Because of the scales that enabled such efficiencies and consistencies of production, fewer people were actually involved in it. Thus, a prevailing irony—as people in the US began to eat more chicken, fewer people had any encounters with living chickens. The trajectory of techno-logical development trends toward greater design complexity, which enables greater sophistication and power in technologies. Fewer people have the technical expertise to contribute to design and manufacture, and thus more people interact with these items only at the level of consumer.

Overall, then, we can see that the requirements that accompany an increase in our ability to assert our intention onto the natural world also contribute, at least in the case of chickens, to an increased separation between producers and consumers, as well as an increase in consumers generally. Where before it may have signified the backyard or the neigh-bor's farm, what it means to grow chickens to sell and to eat becomes reflective of highly specialized knowledge in highly specialized spaces. More to the point, users may be increasingly disconnected from how such objects come to be produced. As chicken becomes meat, it ceases to remain an animal. As we noted in A Chicken, Part 1, this separation is not just conceptual, but also physical. Chickens are moved to the factory, cen-tralized and isolated from human settlements. It is a move partially exac-erbated by the transition to the laboratory.[38]

This disconnection is not unique to chickens, of course, and can have important consequences to the longevity and value of an object to the person who uses it. The philosopher Peter-Paul Verbeek has analyzed this phenomenon of the disposable object that emerges as a result of user dis-connection.[39] He notes that, for users who only relate to the object as a through point to other actions, because the workings of the mechanism remain opaque, such an object is more likely to be seen as disposable and lacking in any other value. When the inner workings of an object are not seen as a mystery to be solved, but rather a complexity to be ignored, it helps to contribute to an object's disposability. There are innumerable ways to be a user of an object. Verbeek argues, from that he terms a post-phenomenological perspective, that objects lose their capacity to be a focus of the user's attention and can instead become conduits for other activity. This kind of pass through contributes to a loss of connection between users and technology, Verbeek argues, especially as those objects become black boxes not meant to be pried open (a state of affairs that can be reinforced by a voided warranty warning if that object is opened or the protective cover removed). This move represents increased complexity on the production side, and results in a further simplification of the user interface and the role of the user.[40]

Verbeek is good at noting that this move toward greater design control and thus separation from the user is not inevitable or a necessary trajectory. He identifies, in fact, designers whose specific goals involve creating technologies that engender a sense of care and even love from their users, in part by designing functionality that requires user participation. Two examples he explores are a heater that has rotating pieces which help direct heat and a computer printer with its workings visible and traceable by the user. As we think about what kind of future we wish to pursue, and what kinds of technologies we might develop to get us there, Verbeek's examples remind us of alternative priorities. They also remind us that knowledge and work need not simply be increasingly concentrated. How such responsibilities are distributed, in the hands of a few complex systems or across a wide range of networks, is a very big question for our future, too.

Before we revisit our chicken-filled possibilities, I want to stress one final element. Our choices about that future are not about whether to assert intention in the world, or whether to leave that world alone. It's not about either a body-less chicken patty or feral chickens. It's about the kind of intention, the amount of control, we seek to assert. I don't want to fall into the trap of proposing opposites. We need to remember that we are not deciding whether to accept responsibility for our future, but what kind of responsibility we want and what kind we can handle. What knowledge, what work, by whom. How much control is desired. How much certainty is required. How much energy is expended and by whom. Those are the types of questions which condition our reality.

The Future as a Laboratory

As I noted at the beginning of this chapter, forecasting the future is a tricky business, more likely to result in error than in insight. But the challenge of imagining the kind of future we want to achieve or to avoid should not keep us from doing so. It should just happen in the context of an awareness of those difficulties and the irreducible uncertainty that attends to the activity. The theoretical reflections on technology above does allow us to avoid focusing too heavily on specifics (will the technology come into being, for example? Will it be a viable alternative to meat? If so, what forms will it take?). That type of forecasting is fraught with peril. As we saw, part of the challenge of technology comes in the difficulty of transforming ideas into reality. That process can lead to failure, or can lead to a modification of our initial goals. In the context of this uncertainty about whether the technology will come into existence in a manner we anticipate, we can still explore why we want to pursue this form of technology in the first place (in what ways does it actually solve the problem to which it aims) and what kinds of new responsibilities the intervention requires. This exploration will allow us to reflect on the version of the future the technology promotes.

The problems that *in vitro* meat sets out to solve are well documented (see A Chicken, Part 1) and worth trying to tackle. They involve the consequences of industrial scale meat production and all of its constituent parts: the animals that form the anchor; the cheap protein that is consumed by a burgeoning population at ever increasing amounts; the impacts of regimes of efficiency and consistency on the bodies of animals and the work of humans. To simply accept this state of affairs, the industrial production of cheap animal protein, as inevitable or natural would be to ignore the very real capacity to change that humans have demonstrated. *In vitro* meat addresses these conditions by offering a way forward technologically, through maintaining a consistent output of animal-based protein but radically transforming the inputs and the conditions of production. By taking greater control over how protein is generated, we can bypass the problems that developed in the industrialized context by shifting that context to the laboratory. In considering the possibility of pursuing this form of technological future, however, we would do well to consider two main aspects. First, we should reflect on the problems associated with control. And second, we should think about our own roles and responsibilities in the context of this approach.

As we industrialized chicken production, we did so through exerting greater and greater control over all facets of the bird's lifecycle. But these controls came with costs and with consequences. The problems are at least in part the result of our efforts to control what we only partially understood and could not keep contained. As we explored earlier, however, this relationship between attempts at control and an incapacity to exert it fully is not particular to chicken—rather, they related to all human engagements with technology. And what we also should pay attention to is that, while *in vitro* meat technologies offer means to avoid some of the more confounding problems associated with industrial-scale, conventional meat production, they do so through attempting to exert even more control on the process. We are attempting to insert even more intention into the means through which animal protein is produced, no longer being compelled to grow inedible substances (like bone and feathers), no longer needing to separate protein from bodies, no longer required to mediate between the animal bodies and their environments. We can control the production and composition and purity of the protein, because we direct each element of that process. Though this effort should lead us to ask: instead of a solution, what if control is part of the problem?

Perfect control is certainly a desirable state of affairs. What better way to improve efficiencies and eliminate uncertainties than to have total control over every aspect of the production of protein. Real philosophical disputes arise about whether or to what extent perfect control is something that humans can achieve over the natural world, over what's real. But there's no denying that *in vitro* meat technologies result from attempts to assert greater amounts of control on the process of producing protein through animal bodies.

Even putting aside the question of whether total control is possible, we would still need to confront the burdens of responsibility that accompany our attempts to assert such control. As we discussed above, such forms of responsibility arise in terms of knowledge and in terms of work. For the transition to *in vitro* meat, that burden becomes even more intense than with industrialized chicken production. We must possess or gain a lot of knowledge that has to be particular and precise. We must perform a lot of work that has to be constant and precise. We don't just care for the body that produces protein. We replace that body with something of our own design—the skin, the circulatory system, the skeleton, and the immune system are simulated within a controlled environment. We do the work of the animal body. The systems we create are the animal body.

There is further to go. Again, it may be worth having the debate about the perfectibility of our knowledge,[41] but there certainly is little question that we are not there yet. Part of the challenge associated with this technique is found in our incomplete knowledge about how to create and sustain animal muscle outside of animal bodies, at least as efficiently as it is done inside of those bodies. Set aside the question about whether we will ultimately be successful in this endeavor, and look at the endeavor itself. This goal suggests that we believe it is possible to know the natural world enough to be able to achieve this result. That the world is knowable in this way, and that the lessons of previous attempts to assert control on that world should not force us to call into question the amount of control, but rather the insufficient knowledge we had at that moment. Had we known more, so it goes, perhaps we could have avoided the problems that emerged.

This assumption involves the kind of narcissism that Heidegger warns about in "The Question Concerning Technology"—that we would look into the world and see only ourselves. Because to say that the world is so completely knowable is to say, in other words, that the world is simply an extension of our own intelligence. And that which escapes our knowledge is a state of affairs that can be overcome and should be greeted with hostility—as a threat to the natural order, rather than an expression of the essence of the natural world. Part of that hostility results from the arrogance of assuming a fully knowable world, and part of it is based on the stakes, as we cannot afford to get things wrong when we build technologies that presume perfectible knowledge, and that seek to control the way the world works. The more control, the more precarious our systems, such that if one thing goes wrong the entire system can be vulnerable. If we do not play the role of skin or immune system well enough, for example, and the protein encounters a contaminant, the entire system can become the contaminant.[42] So, in seeking to create a nature that is only what we want to see in it, we also take on the burden of being right.

Thinking about responsibility also means thinking about us, about the roles that humans play in the process. The requirements of knowing and doing are certainly a part of it. But we also have to confront the question of knowing and doing what? That is, why might we see the urgency of the problem such that we would be willing to shoulder all of that responsibility?

One value of exploring the possibility of *in vitro* meat technologies is that they suggest that things could be different than they are currently. We do not have to assume or simply accept industrialized modes of production and the problems associated with them, but instead we can pursue other modes of producing animal protein. We have options, and those options may be better. But our analysis above should remind us that options like *in vitro* meat are not simple or unqualified improvements. We can opt for those methods but only by accepting increased responsibilities and by being willing to deal with whatever new problems might emerge along the way. This begs the question—why? That is, even if we accept that our current protein production is unsustainable and undesirable, why would we pursue this method that increases control of the process, especially given the requirements that attach to control discussed above? The answer, of course, is us, or at least one dominant version of us.

In vitro meat is desirable because it, like other forms of replacement technology, allows us to continue to behave in the ways we are accustomed, but with less damage. We can perpetuate our current eating habits, and continue to eat the foods of which we've grown fond, while also doing good (or at least less bad). We can eat meat that doesn't pollute the environment in the way it has. We can eat meat that doesn't cause the human health problems we have come to expect. We can eat meat that doesn't end the lives of the animals. We can eat meat. And therein lies the virtue of this approach to the problems of meat—the questions of whether to eat meat or in what context or amount have been overwritten by a technical solution. The process of producing meat changes so that we can remain the same.

There are differences in terms of assuming whether keeping humans constant in their eating habits is one of convenience or necessity. That is, some hold that we should be able to continue eating meat, and this method allows us to do so in less troubling ways.[43] Others view the need to pursue *in vitro* and other alternative forms of meat production out of resignation—seeing humans as incapable of change and thus the technical solution is the only route forward.[44] *Because* we will continue to eat meat, our only solution is to produce meat in a manner that eliminates, or at least minimizes, harm. The technology needs to be better, so we can remain the same or because we *will* remain the same. Any view of the future and what it can look like needs to take seriously the sense of who we are or who would could be. Will we be individuals whose moral choices are obscured by technological processes that seek to make the world better on our behalf, as a way to preserve current behaviors? Or will we be individuals who are willing to confront current behaviors as a way of deciding the kind of technologically mediated future we want to endorse?

Notes

1. Speaking of our current climate-changed world, Johnston notes that, even when things are beyond bleak, we "still have the chance to *make the space* for hope." Johnston, Emily. "Loving a Vanishing World." *Medium* (blog), June 14, 2019. https://medium.com/@enjohnston/loving-a-vanishing-world-ace33c11fe0. The novelist James Bradley also notes that it is important for narratives to explore the "tension between a realistic assessment of the scale of the problem and the desire to preserve some space for change." Quoted in Brady, Amy. "How Will Climate Change Affect Your Grandchildren?" Chicago Review of Books, October 24, 2017. https://chireviewofbooks.com/2017/10/24/burning-worlds-james-bradley-clade-interview/.

2. See, in particular, Fassler, "Lab-grown meat is supposed to be inevitable," which explores the chemical engineer David Humbird's feasibility study of the technology, which suggests that this method may never reach the level of production and scale required to be a reasonable replacement for conventionally grown meat. For examples of previous historical failures, see Belasco, *Meals to Come*.

3. Take, for example, the case of soylent, a product designed to be functionally equivalent to a meal, but widely panned as nigh undrinkable. See, for example, Widdicombe, Lizzie. 2014. "The End of Food," May 5, 2014. https://www.newyorker.com/magazine/2014/05/12/the-end-of-food.

4. Murphy does argue that apocalyptic narratives do contain elements of utopian visioning, in terms of what should have happened to avoid the apocalypse that befell the imaginary future. Murphy, Amy. "Nothing Like New: Our Post-Apocalyptic Imagination as Utopian Desire." *Journal of Architectural Education* 67, no. 2 (July 3, 2013): 234–42. https://doi.org/10.1080/10464883.2013.817166.

5. Schulz, Kathryn. 2015. "The Earthquake That Will Devastate the Pacific Northwest." *The New Yorker*, July 13, 2015. https://www.newyorker.com/magazine/2015/07/20/the-really-big-one.

6. Issues related to not labelling things because they are functionally equivalent and thus would raise the specter of concern when there need not be any. See Keim, "FDA: Don't Ask, Don't Tell…", and Tanne, Janice Hopkins. 2008. "FDA Approves Use of Cloned Animals for Food." *BMJ : British Medical Journal* 336 (7637): 176. https://doi.org/10.1136/bmj.39468.528368.DB.

7. All things are not, of course, equal. Many consumers do not have the income that would allow them to reflect their values fully in the marketplace.

8. Vogel, *Thinking like a Mall*.

9. Preston, Christopher J. 2019. *The Synthetic Age: Outdesigning Evolution, Resurrecting Species, and Reengineering Our World*. Reprint edition. Cambridge, Massachusetts: MIT Press, p. 173.

10. Vogel, *Thinking like a Mall*, 122.

11. Vogel, *Thinking like a Mall*, 140.

12. Jenny Price puts the dilemma this way: the question is not whether we change the environment, but rather "How can we *change* environments vastly better?" See Price, Jenny. *Stop Saving the Planet!: An Environmentalist Manifesto*. W. W. Norton & Company, 2021, 10.

13. Heidegger, Martin. 1977. *The Question Concerning Technology, and Other Essays*. New York: Harper & Row.

14. Vogel, *Thinking like a Mall*, 144, emphasis in the original.

15. Latour, Bruno. 2011. "Love Your Monsters: Why We Must Care for Our Technologies As We Do Our Children." Edited by Ted Nordhaus and Michael Shellenberger. *Breakthrough Journal*, no. 2 (Fall).

16. Thoreau, Henry D. 1996. *Faith in a Seed: The Dispersion Of Seeds And Other Late Natural History Writings*. Edited by Bradley P. Dean. Reissue edition. Island Press.

17. See Pollan, Michael. *Second Nature: A Gardener's Education*. New York, NY: Grove Press, 2003. Here, Pollan explores the requirements of work that come with maintaining a garden, which is a clear effort to impose one's intentions on the natural world.

18. I started growing garlic as my primary crop, which succeeded despite my failures.

19. Faith, that is, in how a seed works. We may need faith in our awareness of the seed's wider impacts on the surrounding ecologies.

20. Latour, "Morality and Technology."

21. Latour, "Morality and Technology," 252. Emphasis in original.

22. Adorno, Theodor W. *Negative Dialectics*. Translated by E.B. Ashton. New York: Continuum, 1983.

23. Vogel, *Thinking Like a Mall*, 113.

24. This example just focuses on the use of the hammer. Even more work could be reckoned up if we analyzed the very presence of the hammer itself, as before I could ever use it, the tool had to be made. For the head of the hammer itself, the metal had to be mined, refined, and shaped. Someone had to forge it. The presence of that kind of work within objects—the kinds of efforts and networks necessary to make that object present to the user—is what Bruno Latour calls a fold. See Latour, "Morality and Technology."

25. McPhee, John. 1990. *The Control of Nature*. Reprint edition. New York: Farrar, Straus and Giroux.

26. Pun intended.

27. "Too far" here is defined in two ways—safety vis-à-vis the road, and propriety vis-à-vis my neighbors' property and their desire not to encounter the unexpected chicken poop.

28. Some people who raise chickens increase the utility of a fence by clipping the wings of their birds.

29. Debeaking, or more euphemistically beak-trimming, chickens.

30. Locke, John. 1690. "Second Treatise of Civil Government." 1690. https://www.marxists.org/reference/subject/politics/locke/ch05.htm, Ch. 5.

31. Berry, "The Pleasures of Eating."

32. Pollan, Michael. 2002. "Power Steer." *New York Times*, March 31, 2002, sec.Magazine.http://www.nytimes.com/2002/03/31/magazine/power-steer.html?pagewanted=all&src=pm.

33. I have, of course, improvised a hammer out of necessity. I feel confident that I am not alone in that fact, either.

34. Thwaites, Thomas. 2011. *The Toaster Project: Or a Heroic Attempt to Build a Simple Electric Appliance from Scratch*. 1st Edition. New York: Princeton Architectural Press. For a satirical take on the trend of making one's own staples, see https://www.theonion.com/grueling-household-tasks-of-19th-century-enjoyed-by-sub-1819565927.

35. See Striffler, *Chicken*.

36. See, for example, Alexander, *The $64 Tomato*.

37. Fitzgerald, Deborah K. 2003. *Every Farm a Factory: The Industrial Ideal in American Agriculture*. New Haven: Yale University Press.

38. The dream of *in vitro* meat, were it to come to fruition, would involve the migration of production from the lab to the kitchen counter, in the guise of a "meat machine" analogous to a bread machine. Even if this came to pass years from now, the workings of that machine would likely be opaque to the common user, a case where physical proximity does not entail closer understanding.

39. Verbeek, Peter Paul. 2005. *What Things Do: Philosophical Reflections on Technology, Agency, and Design*. University Park, PA: Pennsylvania State University Press.

40. As Tannen puts it, in reference to user interface design, "making something 'simpler' is often a case of relocating complexity, rather than eliminating it from the user-technology relationship"—in this case, relocating it within the production system and away from the perspective of the user (Tannen, "Simplicity"). See also Norman, Donald A. 2010. *Living with Complexity*. Cambridge, MA: The MIT Press.

41. This is something Tannert, et al., refer to as ontological uncertainty (Tannert, Christof, Horst-Dietrich Elvers, and Burkhard Jandrig. 2007. "The Ethics of Uncertainty. In the Light of Possible Dangers, Research Becomes a Moral Duty." *EMBO Reports* 8 (10): 892–96. https://doi.org/10.1038/sj.embor.7401072).

42. See Specter "Test-Tube Burgers," 37. This is also an example of the new kinds of accidents that are coextensive with new technologies. See Loeb, "Inventing the Shipwreck."

43. See Hopkins & Dacey, "Vegetarian Meat."

44. See Deych, "How One Vegan Views In-Vitro Meat."

BIBLIOGRAPHY

Adorno, Theodor W. *Negative Dialectics*. Translated by E.B. Ashton. New York: Continuum, 1983.

Alexander, William. *The $64 Tomato: How One Man Nearly Lost His Sanity, Spent a Fortune, and Endured an Existential Crisis in the Quest for the Perfect Garden.* Chapel Hill, NC: Algonquin Books of Chapel Hill, 2007.

Belasco, Warren J. *Meals to Come: A History of the Future of Food.* Berkeley: University of California Press, 2006.

Berry, Wendell. "The Pleasures of Eating." ecoliteracy.org, June 29, 2009. https://www.ecoliteracy.org/article/wendell-berry-pleasures-eating.

Brady, Amy. "How Will Climate Change Affect Your Grandchildren?" Chicago Review of Books, October 24, 2017. https://chireviewofbooks.com/2017/10/24/burning-worlds-james-bradley-clade-interview/.

Deych, Rina. "How One Vegan Views In-Vitro Meat," 2005. http://www.rrrina.com/invitro_meat.htm.

Fassler, Joe. "Lab-grown meat is supposed to be inevitable. The science tells a different story." *The Counter*, September 22, 2021. https://thecounter.org/lab-grown-cultivated-meat-cost-at-scale/.

Fitzgerald, Deborah K. *Every Farm a Factory: The Industrial Ideal in American Agriculture.* New Haven: Yale University Press, 2003.

Heidegger, Martin. *The Question Concerning Technology, and Other Essays.* New York: Harper & Row, 1977.

Hopkins, Patrick D., and Austin Dacey. "Vegetarian Meat: Could Technology Save Animals and Satisfy Meat Eaters?" *Journal of Agriculture and Environmental Ethics* 21 (2008): 579–96.

Johnston, Emily. "Loving a Vanishing World." *Medium* (blog), June 14, 2019. https://medium.com/@enjohnston/loving-a-vanishing-world-ace33c11fe0.

Keim, Brandon. "FDA: Don't Ask, Don't Tell on Cloned Meat." *Wired*, January 15, 2008. https://www.wired.com/2008/01/fda-dont-ask-do/.

Latour, Bruno. "Love Your Monsters: Why We Must Care for Our Technologies As We Do Our Children." Edited by Ted Nordhaus and Michael Shellenberger. *Breakthrough Journal*, no. 2 (Fall 2011). http://thebreakthrough.org/index.php/journal/past-issues/issue-2/love-your-monsters.

———. "Morality and Technology: The End of the Means." *Theory, Culture, & Society* 19, no. 5/6 (2002): 247–60.

Lee, Jonathan. *High Dive*. New York: Vintage Books, 2017.

Locke, John. "Second Treatise of Civil Government," 1690. https://www.marxists.org/reference/subject/politics/locke/ch05.htm.

Loeb, Zachary. "Inventing the Shipwreck." *Real Life*. January 3, 2022. https://reallifemag.com/inventing-the-shipwreck/.

McPhee, John. *The Control of Nature*. Reprint edition. New York: Farrar, Straus and Giroux, 1990.

Murphy, Amy. "Nothing Like New: Our Post-Apocalyptic Imagination as Utopian Desire." *Journal of Architectural Education* 67, no. 2 (July 3, 2013): 234–42. https://doi.org/10.1080/10464883.2013.817166.

Norman, Donald A. *Living with Complexity*. Cambridge, Mass: The MIT Press, 2010.

Pollan, Michael. "Power Steer." *New York Times*, March 31, 2002, sec. Magazine. http://www.nytimes.com/2002/03/31/magazine/power-steer.html?pagew anted=all&src=pm.

———. *Second Nature: A Gardener's Education*. New York, NY: Grove Press, 2003.

Preston, Christopher J. *The Synthetic Age: Outdesigning Evolution, Resurrecting Species, and Reengineering Our World*. Reprint edition. Cambridge, Massachusetts: MIT Press, 2019.

Price, Jenny. *Stop Saving the Planet!: An Environmentalist Manifesto*. W. W. Norton & Company, 2021.

Schulz, Kathryn. "The Earthquake That Will Devastate the Pacific Northwest." *The New Yorker*, July 13, 2015. https://www.newyorker.com/magazine/2015/07/20/the-really-big-one.

Specter, Michael. "Test-Tube Burgers." *The New Yorker*, May 23, 2011.

Tanne, Janice Hopkins. "FDA Approves Use of Cloned Animals for Food." *BMJ : British Medical Journal* 336, no. 7637 (January 26, 2008): 176. https://doi.org/10.1136/bmj.39468.528368.DB.

Tannen, Rob. "Simplicity: The Distribution of Complexity." Boxes and Arrows, January 30, 2007. https://boxesandarrows.com/simplicity-the-distribution-of-complexity/.

Tannert, Christof, Horst-Dietrich Elvers, and Burkhard Jandrig. "The Ethics of Uncertainty. In the Light of Possible Dangers, Research Becomes a Moral Duty." *EMBO Reports* 8, no. 10 (October 2007): 892–96. https://doi.org/10.1038/sj.embor.7401072.

Thoreau, Henry D., Robert Richardson, and Gary Paul Nabhan. *Faith in a Seed: The Dispersion Of Seeds And Other Late Natural History Writings*. Edited by Bradley P. Dean. Reissue edition. Island Press, 1996.

Verbeek, Peter Paul. *What Things Do: Philosophical Reflections on Technology, Agency, and Design*. University Park, PA: Pennsylvania State University Press, 2005.

Vogel, Steven. *Thinking like a Mall: Environmental Philosophy after the End of Nature*. 1 edition. The MIT Press, 2015.

Widdicombe, Lizzie. "The End of Food," May 5, 2014. https://www.newyorker.com/magazine/2014/05/12/the-end-of-food.

CHAPTER 8

A Conclusion: Protein Machines, the Human Kind

Abstract The book concludes with a more focused look at one part of the protein machine-humans. The value of taking an historically informed, parallel look at two different means of making chicken is that it allows us to see that humans have played various roles in both systems. More to the point, when contemplating the type of future we want to pursue-one that happens because of us, not just to us-we would be wise to remember that humans have changed, can change, and will change in our relationships to the protein machines of the future.

Keywords Change • Future • Human • Nature • Technology

> *To become a maker is to make the world for others, not only the material world but the world of ideas that rules over the material world, the dreams we dream and inhabit together.*
> —Rebecca Solnit, *The Faraway Nearby*, Reprint edition, New York: Penguin Books, 2014

This book has ostensibly focused on chickens and the increasingly complex machine systems that surround them. This story is also, and increasingly, about us—about humans and the tensions we experience between fitting ourselves to the world and fitting the world around ourselves. Both

W. Galusky, *Protein Machines, Technology, and the Nature of the Future*, https://doi.org/10.1007/978-3-031-08717-2_8

happen in our interactions with technologies and technological systems. The choice is not whether to go about this process of co-production, but rather how—what priorities and what values will we attempt to express? One possible way to engage the natural world, out of many, involves *in vitro* meat. The goals of this technological pursuit are consistent with a very specific way of being in and knowing nature and of knowing ourselves. It assumes that the natural world can be controlled to reflect more fully human desires, and the process of requiring nature to express itself that way and of requiring ourselves to conform to those requirements is an improvement over the current context. In this case, the desire to consume the meat we are used to without needing to kill animals. It suggests, from the position of the consumer, at least, a posture of innocence toward the world—a way to have our chickens and eat them, too. There is also the desire to precisely control the composition of what we eat. From the point of view of the producer, this approach to meat is a way to render the world completely legible to human forms of understanding. It presumes that humans can know the processes (how nature works) and the products (what nature is) to create exactly what we want. This lab-based future is an extension of the industrialized protein machine built to provide chicken at ever increasing quantities and ever dwindling market prices; the innovations are meant to correct the more egregious problems and uncertainties of the factory.

An important part of the book's goal has been to place the development of *in vitro* within a historically consistent trajectory—one that moves the U.S. consumer toward simpler and simpler versions of chicken and further and further away from chickens. This simplicity is deceptive, in that it is accomplished by increasingly complex knowledge sets and processes meant to guide or replace activities more typically done by the animal itself. But those complexities are easy to lose sight of, to our detriment. Importantly, we might also lose sight of the complexities associated with ourselves. That's where we left off the last chapter, thinking about a kind of future and our place in it. And our vision for the future begs a question—are we fixed? Must we remain in our current iteration—as consumers of cheap, consistent animal protein? I think the lessons we can take from our analyses throughout this book are the following: We have changed. We will change. And, I would assert, we can change.

We Have Changed

It's important to remember that we have only recently become a nation of chicken eaters. One advantage to taking an historical view of the desire for chicken protein is that it demonstrates how malleable human desire for chicken has been, at least in the United States. If we expanded our historical scope, we could even argue that widespread choice about what to eat from a bevy of options is a relatively recent phenomenon. The presence of an abundance of cheap, available food for a large portion of the populace[1] means that our choices about what to eat become a means by which we can express our identities and reflect our values.[2] In fact, our capacity to eat a variety of foods, coupled with a plethora of food choices, has created their own dilemmas as we try to figure out what we should eat, and how we might prioritize health, or pleasure, or community. This state of affairs can lead to confusion and some unfortunate meals—I still feel scarred as a child of the 1980s, growing up in a time when fat became enemy number one to a subset of eaters, and being compelled to eat a homemade pizza made with fat-free cheese. It was a cheese that didn't so much melt as leak whatever counted as oil held within its three-dimensional "shredded" shape. This landscape of food choices can create as much anxiety as empowerment, given the uncertainty of what we should eat from all the options.[3]

Thus, the notion that we are somehow fixed eaters who won't change our eating habits in response to certain trends or necessities appears belied by our recent past. So it seems odd that we would invest so heavily in the idea that human meat eating is fixed to the point of being a requirement. From the point of view of technology and responsibility that we explored in the previous chapter, it seems ironic that there would be so much willingness to take the challenging technological route and see it as easier than getting people to consider the implications of their own meat eating, as individuals or as a body politic. Again, history shows us that we are not fixed, and can make different choices in the face of shifting contexts.

We didn't just change what or how we ate. A major focus of our analysis has been to unpack as much of the technological systems as possible—not just the thing, or in this case the chicken. Instead, we want also to reckon up what makes that chicken possible as such, and in turn how the presence of that chicken allows us to configure experiences. So, to confront the kind of changes that have occurred historically in our relationship to chicken, it's important not to stop at the point of eating more of it. We should also recognize the extent to which the production processes moved out of sight and out of the range of everyday experience for most people. Chickens became chicken. They went from being an animal one might raise as a kind of food security or supplemental income to, for most people, something one might encounter prepackaged in a supermarket or already fried for us at the counter of a fast-food restaurant. Chicken became meat, and so we tended to relate to the protein rather than the animal. This occurred with other animals, as well.

The production of protein shifted to the periphery of most people's experience, through the concentration and industrialization of the processes. As such, we changed how we related to chickens and the process of chicken production.[4] It became easier to ignore the agricultural facts of protein production—that there were animals involved, and farmers, and processors, and spaces and ecologies and communities—if they were not ours. We became primarily consumers. And one could argue that this sense of invisibility facilitated by distance is part of the problem associated with modern protein production. Or, at least, it helped to exacerbate the problems because there was a disconnect between what people ate and how it was produced, creating incentives and pressures to produce food as cheaply and consistently as possible. The prevailing sentiment being that meat should be cheap and consistent, regardless of what that means in terms of production.

We have changed. What we eat. How we eat it. Why we eat it.

We Will Change

That history also should keep us honest in assessing what might happen in the future. The historical fact of human change as the result in innovations in protein production (both in what we ate and in how we related to the processes of producing that food) serves as an important corrective against certain assumptions. In particular, the assumption that *in vitro* meat technologies will simply insert themselves into the existing cultural landscape and human identity and have no effect other than how chicken arrives on our plates. Our understanding of technology should lead us to the conclusion that many aspects of food, of animals, of ourselves, and of our understanding of nature will be altered. There will be increased invisibility and/ or mystery regarding how the protein is made. Increased distance between us and the animals that we normally associated with the protein. Increased choice (and the anxieties that attach to choosing) with regard to what we eat, given that so much more is under human direction in terms of composition. And an increased sense that nature can be simply made to do our bidding, suggesting that failures are not accidents but negligence.[5]

It would be impossible to describe the precise contours of those changes, given the uncertainties that remain in terms of what, if any, particular shape *in vitro* meat production will take. For example, we would confront different technologies if *in vitro* meat production capabilities moved into the home, in the form of a bread-machine model, or if production was centralized and the protein showed up in the supermarket alongside other, more conventional products. Even if a dominant mode emerged, however, we would still face difficulties in predicting how it might change human action and interpretation—there is an equally compelling human history of people using devices in ways their designers never intended.[6] But we can assume that humans will change in response to this new technological possibility, and thus catering to some fixed human essence again appears a fool's errand. The narratives we explored in the last chapter help to spell out this interactivity and co-produced relationship. They portray worlds that contain humans and technologies adapted to each other. We don't have to take them as prognosticators or predictive of how the world will be (and it would be hopeful of us NOT to do so). But they remind us that the future is refracted through this interplay of technological possibility and human reality.

We Can Change

So now we have dealt with the past and with the future—both of which testify to the human tendency to change through our interaction with technological systems. That change can be subtle and tangential, existing at the periphery of our experience. We may not be aware of how much our goals change in response to the detours we need to take in order to achieve a goal, or how our goals subtly shift in response to abilities. That change can also be intentional and directed—because of a perceived increase in disconnection between the food we eat and the processes that make such food possible, we alter our everyday habits. We have backyard chickens. We stop eating certain products. We grow our own food or get to know the people who do. So what about us at present, confronting an array of possible futures? To what extent can we change, and be intentional about it? To what extent should we change, instead of asking the natural world to change? Who should accommodate whom?

One important thing to consider when contemplating this potential *in vitro* meat future is the kind of changes we are prepared to make, and who is asked to make them. If we consider the problems of animal-based protein production a purely technological one, and hold to the notion that technologies are mere means to achieve human ends, then the *in vitro* approach makes perfect and even inevitable sense. What better way to deal with the problems of a previous technology then to produce a better, more efficient technology. We can have nature (in the guise of chickens, and chicken, and cells, and processes) better accommodate our desires by asserting more control over what nature can be and do.

This technological effort focuses on making the question of whether to eat meat a non-moral question for the consumer. One could eat meat and not contribute to the killing of the animal from which the protein derives, but the point would be from the point of view of the consumer to just continue eating meat as before—to change nothing about diet. To do good by doing nothing different. Assuming the choice is visible (in terms of branding or labeling or any other marker of distinction), for the individual consumer it's one of substitution. Which form of these indistinguishable proteins will I choose? At least, it's meant to make the moral question easy and entail no sacrifice. Our role in this machine is simple. It doesn't ask much of us.

This represents the strategy of products like Impossible Foods and Beyond Foods—making meat morally palatable by making it out of plants, but otherwise indistinguishable from other forms of protein. One of the slogans for Beyond Meat is "Eat What You Love".[7] For Impossible Foods, the mission is: "To Save Meat. And Earth."[8] They go on:

> Using animals to make meat is a prehistoric and destructive technology. We're making meat from plants so that we never have to use animals again. That way, we can eat all the meat we want, for as long as we want. And save the best planet in the known universe.[9]

These products play a similar game—creating a reasonable facsimile, to enable easy choices and better choices. Doing well by doing nothing different. Morality that is either easy or sneaky—something that is done for us or to us.

It may be that this is the route we choose. But to do so, to turn questions of moral and social values into technical questions of substitution and simulation is not to make those moral questions disappear. It's to shift them, to allow others to make them on our behalf. It is to accept a simpli-fied role for ourselves in the functioning of this protein machine. And if we think through this technological intervention in the context of our analy-sis, which complicates the neutrality of tools and acknowledges the burden of responsibilities attached to our efforts to assert control, then we have a different kind of decision to make about our future and our present. One prime issue relates to the irony attached to the *in vitro* approach. That is, on the one hand, humans are collectively seen as relatively fixed in terms of our diets, desires, and behavior. We are consumers of animal protein, and so we need to produce protein better to accommodate those desires. On the other hand, we also see ourselves of having the capacity to assume such new burdens of responsibility in order to make this technological approach possible. Of course, the "we" in each sentence does not have exactly the same referent—the burdens of responsibility might be shared by all of us, but are directed by a much smaller number of technical experts. For those who do not possess the expertise to direct and manage these processes, we might feel exonerated. We also position ourselves as passive consumers of whatever is produced on our behalf.

But the responsibilities we assume are not merely technical. They are *also* ethical and social. Thus, another approach to the problems associated with industrialized production is not just to make our technique better, but to also strive to make ourselves better. That is, we can think through the problems of meat by making different collective choices. Here's where we should take up the issues of inconsequentialism again. As we discussed in the previous chapter, this phenomenon is associated with individual choices and actions, particularly consumer choices and actions, in response to complex social problems. Vogel's focus here is on climate change, but can be extended to other complex, large-scale problems. In this case, not only is the individual decision, for example, to cut emissions by not driving likely to have no discernable effect, it can also be seen as irrational (the individual bears increased costs in technology and time compared to oth-ers while making no tangible difference). Vogel wants us to consider the true nature of the problem, namely:

> The problem isn't that the effect of one's [individual] actions on a social aggregate is inconsequential; that's true more or less by definition. The problem is that one's individually 'inconsequential' actions to cut emissions can only have a consequential effect, and hence count as morally valuable

'contributions,' if they are coordinated with the (again individually inconse-
quential) actions of many others.[10]

We must think of ourselves as part of a collective agency, as a group of
people who can work in common to, in Vogel's words, "*build the sort of
community* capable of"[11] solving the problems we are trying to avert.

This approach to the issue of change leads us to another paradox, however, because technologies already are collective acts, in a sense. Were I to
purchase *in vitro* meat at the store, it would feel like an individual consumer act alone, but it does connect me to all of those processes and
people we met above.[12] Despite the truth of this interconnectedness, the
individual consumer doesn't have to directly confront those complexities
and responsibilities, may not know how to, and instead can feel alienated
from that technological system. Vogel notes that this alienation emerges
because people operate in a kind of bubble that obscures the connections
between collective human action and large-scale, wicked phenomena like
climate change or the market. So, these forces appear to be external, non-
human, and in some ways "natural." Alien from us in terms of feeling
disconnected to our actions. He calls for a twofold recognition—to understand that these phenomena are the product of human activities that have
been aggregated beyond the scale of individual action (though still the
product of those actions), and thus to recognize the limits of individuals
within that frame of reference and the need to act collectively. Latour,
discussing different forms of globalism, discusses the challenges associated
with this sort of reckoning, which can cause us to "complain in general,
and [succumb to] the impression of no longer having any leverage that
could enable us to modify the situation."[13] This feeling of disempower-
ment results in part in how daunting it is to try and account for how our
world is made, and what our parts are in it.

We can, however, see our role in the shaping of the future. We would
need to do so by remembering that to act through technology is to act
collectively. In a way, that's exactly what the researchers for *in vitro* meat
are trying to do—act in a way that shapes the collective. But we can expand
the people who are involved in the process, who participate in the active
shaping of that collective, and can do so in a way that does not require
such huge burdens of technological responsibility. We would do so by seeing ourselves as active members of a collective, rather than the passive,
alienated outsiders who are merely acted upon.

One way to do so is to recognize that we can be moral beings. We can be encouraged to make different choices in the face of changing contexts and awareness. We can be asked to be responsible not for the newly designed and configured bodies of animals in order to consume guilt-free protein. But, instead, or at least in addition, we can be responsible for reflecting on our moral place in the world and the kinds of actions that are or might be required of us. The problems are social and ethical, not just technical. In fact, we can't disentangle them. The future is our future, and is not just about technologies, but also and foremost about us.

Notes

1. Though certainly not everyone. For example, according to the USDA, 11.1% of US households experienced food insecurity in 2018, which amounts to roughly 14.4 million households ("USDA ERS – Food Security and Nutrition Assistance." n.d. Accessed January 16, 2020. https://www.ers.usda.gov/data-products/ag-and-food-statistics-charting-the-essentials/food-security-and-nutrition-assistance/).

2. One place where the depth of this attention to values and identity as related to chickens is satirized is in the opening episode of *Portlandia*, where diners can interrogate the entire life history of their meal, Colin the Chicken (*Colin the Chicken | Portlandia | IFC*. n.d. Accessed January 14, 2020. https://www.youtube.com/watch?v=G__PVLB8Nm4). When you tell people you write about chickens, this sketch gets referenced a lot. *A lot.*

3. See Pollan, *The Omnivore's Dilemma*, and Pollan, Michael. 2008. *In Defense of Food: An Eater's Manifesto*. New York: The Penguin Press. He's made a kind of industry of this.

4. See Bulliet, Richard W. 2002. "Human-Animal Relations in the Era of Postdomesticity." 2002. http://www.fathom.com/feature/35184/index.html.

5. McPhee, *The Control of Nature*. This happens with old river control mechanism. When flooding did occur downstream, or fishing wasn't optimal, people blamed the Corps rather than their bad luck. No more acts of god.

6. Famously, the telephone.

7. "Recipes." n.d. *Beyond Meat – The Future of Protein*™ (blog). Accessed January 21, 2020. https://www.beyondmeat.com/recipes/. Ingredients for the Beyond Burger: Water, Pea Protein*, Expeller-Pressed Canola Oil, Refined Coconut Oil, Rice Protein, Natural Flavors, Cocoa Butter, Mung

Bean Protein, Methylcellulose, Potato Starch, Apple Extract, Pomegranate Extract, Salt, Potassium Chloride, Vinegar, Lemon Juice Concentrate, Sunflower Lecithin, Beet Juice Extract (for color).

8. "Mission – Impossible Foods." n.d. Accessed January 21, 2020. https://impossiblefoods.com/mission/.

9. "Mission – Impossible Foods." The ingredients: Water, Soy Protein Concentrate, Coconut Oil, Sunflower Oil, Natural Flavors, 2% or less of: Potato Protein, Methylcellulose, Yeast Extract, Cultured Dextrose, Food Starch Modified, Soy Leghemoglobin, Salt, Soy Protein Isolate, Mixed Tocopherols (Vitamin E), Zinc Gluconate, Thiamine Hydrochloride (Vitamin B1), Sodium Ascorbate (Vitamin C), Niacin, Pyridoxine Hydrochloride (Vitamin B6), Riboflavin (Vitamin B2), Vitamin B12.

10. Vogel, *Thinking like a Mall*, 215.

11. Vogel, *Thinking like a Mall*, 214, emphasis in original.

12. As Wendell Berry reminds us, "eating is an agricultural act." Berry, "The Pleasures of Eating."

13. Latour, Bruno. 2018. *Down to Earth: Politics in the New Climatic Regime.* Polity Press, p. 96.

BIBLIOGRAPHY

Berry, Wendell. "The Pleasures of Eating." ecoliteracy.org, June 29, 2009. https://www.ecoliteracy.org/article/wendell-berry-pleasures-eating.

Beyond Meat. "Recipes." *Beyond Meat – The Future of Protein*™ (blog). Accessed January 21, 2020. https://www.beyondmeat.com/recipes/.

Bulliet, Richard W. "Human-Animal Relations in the Era of Postdomesticity," 2002. http://www.fathom.com/feature/35184/index.html.

Colin the Chicken | Portlandia | IFC. Accessed January 14, 2020. https://www.youtube.com/watch?v=G__PVLB8Nm4.

Impossible Foods. "Mission." Accessed January 21, 2020. https://impossible-foods.com/mission/.

Latour, Bruno. *Down to Earth: Politics in the New Climatic Regime.* Polity Press, 2018.

McPhee, John. *The Control of Nature.* Reprint edition. New York: Farrar, Straus and Giroux, 1990.

Pollan, Michael. *In Defense of Food: An Eater's Manifesto.* New York: The Penguin Press, 2008.

———. *The Omnivore's Dilemma: A Natural History of Four Meals.* New York: Penguin Press, 2006.

Solnit, Rebecca. *The Faraway Nearby*. Reprint edition. New York, New York: Penguin Books, 2014.

USDA Economic Research Service. "Food Security and Nutrition Assistance." Accessed January 16, 2020. https://www.ers.usda.gov/data-products/ag-and-food-statistics-charting-the-essentials/food-security-and-nutrition-assistance/.

Vogel, Steven. *Thinking like a Mall: Environmental Philosophy after the End of Nature*. 1 edition. The MIT Press, 2015.

INDEX

© The Author(s), under exclusive license to Springer Nature
Switzerland AG 2022
W. Galusky, *Protein Machines, Technology, and the Nature of the
Future*, https://doi.org/10.1007/978-3-031-08717-2